The science and practice of welding

of welding

VOLUME 1

Welding science and technology

A.C.DAVIES

B.Sc (London Hons. and Liverpool), C.Eng., M.I.E.E., Fellow of the Welding Institute

EIGHTH EDITION

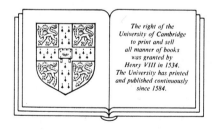

The right of the
University of Cambridge
to print and sell
all manner of books
was granted by
Henry VIII in 1534.
The University has printed
and published continuously
since 1584.

CAMBRIDGE UNIVERSITY PRESS

Cambridge

London   New York   New Rochelle

Melbourne   Sydney

Published by the Press Syndicate of the University of Cambridge
The Pitt Building, Trumpington Street, Cambridge CB2 1RP
32 East 57th Street, New York NY 10022, USA
10 Stamford Road, Oakleigh, Melbourne 3166, Australia

First published 1941
Second edition 1943
Third edition 1945
Reprinted 1947, 1950
Fourth edition 1955
Reprinted 1959
Fifth edition 1963
Reprinted 1966, 1969, 1971
Sixth edition 1972
Reprinted 1975
Seventh edition 1977
Reprinted with revisions 1981
Eighth edition 1984
Reprinted with new Appendix 3 1986

Printed in Great Britain at the University Press, Cambridge

Library of Congress card number 83-23225

British Library cataloguing in publication data

Davies, A. C.
  The science and practice of welding. – 8th ed.
  Vol. 1: Welding science and technology
  1. Welding
  I. Title
  671.5′2   TS227

ISBN 0 521 26113 9 hard covers
ISBN 0 521 27839 2 paperback
(ISBN 0 521 21557 9 7th edition)

621·977 D

Contents

Preface to the eighth edition

The science and practice of welding has now been divided into two volumes. Volume 1 is *Welding science and technology* and Volume 2 is the *Practice of welding*.

This Volume 1 includes the physics, chemistry, metallurgy, equilibrium diagrams, basic electrical principles, testing of welds (non-destructive and destructive), drawing and welding symbols. An appendix with tables and conversions, etc. is also included.

The text has been brought up to date where required and there are new sections on aluminium and its alloys incorporating the new classifications, magnesium and its alloys, and steel and steel alloys with examples of steel classification.

As much accentuation is placed on the non-destructive testing (NDT) of welds this section has been enlarged with new sections on penetrant dye, X-ray and gamma-ray testing.

My thanks are due to each of the following firms who have rendered every technical assistance and have supplied information and photographs as indicated.

Air Products Ltd: NDT of welding fabrications and all radiographs.

Alcan Wire Ltd: classification and welding of aluminium and its alloys.

Aluminium Federation: classification and properties of aluminium and its alloys.

Amersham International: X-ray and gamma-ray testing of welds and gamma-ray sources.

Andrex Products (NDT) Ltd: sources of X-rays and the testing of welds with illustrations.

British Oxygen Co. Ltd (Gases Division): gas technology.

Baugh and Weedon Ltd: ultrasonic testing of welds and photographs.

Copper Development Association: classification of copper and its alloys.

Magnesium Elektron Ltd: classification and properties of magnesium and its alloys.

Murex Welding Products: thyristor control of welding sources.

Radiographic Supplies Ltd: diagrams and information on magnetic particle, ultrasonic and dye penetrant testing of welds.

Salford Electrical Instruments Ltd: link testing ammeter.

Testrade Ltd: information and photographs of gamma-ray testing of welds.

The Welding Institute: information on crack tip opening displacement testing.

G. J. Wogan Ltd: information and testing of welds with photographs.

My thanks are also due to Messrs D. G. J. Brunt, A. Ellis, C. Owen, M. S. Wilson and R. A. Wilson for technical help received in the preparation of this edition, and to the City and Guilds of London Institute for permission to include some up to date questions in welding science and technology to add to the questions at the end of the book.

Abstracts of British Standards brought up to date are included by permission of the Sales Department, British Standards Institution, Linford Wood, Milton Keynes MK14 6LE, tel: 0908 320033, from whom copies of the latest complete standards can be obtained.

Oswestry
1984   A. C. Davies

Preface to the seventh edition

The book has been extensively revised with new sections on submerged arc, stud (arc and capacitor), explosive and gravity processes. A new chapter has been added on resistance welding and many sections have been brought up to date including the new processes in iron and steel production, and additional information is included in the chapters on TIG and MIG processes. The new electrode classification to BS 639 1976 has been included together with impact machines and testing.

My thanks are due to each and all of the following firms who have helped me in every way by offering advice and supplying information and photographs as indicated.

A.I. Welders Ltd: flash butt welding technology and photographs.

Air Products: Cryogenic Division, the welding of aluminium alloys and stainless steel; Gases Division, details of helium mixtures.

Avery-Denison Ltd: impact testing machines and photographs.

British Oxygen Co. Ltd: oxy-acetylene welding equipment and photographs, industrial gases and diagrams, manual metal arc welding electrodes and filler wires, manual metal arc, TIG, MIG and plasma welding plant and plasma cutting with diagrams and photographs.

Copper Development Association: the welding of copper and its alloys.

Crompton Parkinson Ltd: stud welding with diagrams.

ESAB Ltd: manual metal arc welding electrodes and filler wires, manual metal arc, TIG, MIG, submerged arc, gravity, and electroslag welding equipment, positioners, and robot welding with illustrations and photographs.

G.K.N. Lincoln Ltd: submerged arc welding equipment, wire electrodes and fluxes.

British Railways Board: details of flash butt and thermit welding of rails, and diagrams of 'adjustment switch'.

British Steel Corporation, Library and Information Services of the Sheffield Laboratories: modern blast furnaces, direct reduction of iron ores, basic oxygen steel, electric arc steel with illustrations.

KSM Stud Welding Ltd: stud welding with diagrams.

Pirelli General Cable Co.: welding cables.

The Welding Institute: information on the classification of electrodes.

Cooperheat Ltd: pre- and post-heating equipment with photographs.

Sciaky Ltd: spot, seam projection and other types of resistance welding with photographs, laser beam welding with photograph.

Henry Wiggin Ltd: the welding of nickel and nickel alloys.

Union Carbide UK Ltd: TIG and MIG technology, plasma welding, cutting and surfacing technology with diagrams.

Yorkshire Imperial Metals Ltd: explosive welding with diagrams.

Birlec Ltd: induction furnace photograph.

Rockwell Ltd: photographs of CO_2 welding equipment.

Gamma-Rays Ltd: information on non-destructive testing with photographs of radiographic equipment.

I would again like to express my thanks to the City and Guilds of London Institute for permission to reproduce, with some amendments, examination questions set in recent years and to Mr D. G. J. Brunt, T. ENG (C. E.I), F.I.T.E., ASSOC. MEM.I.E.E., M.WELD.I., for help in the reading of proofs, and to Mr M. S. Wilson, B.SC., M.MET., for help in the revised sections in metallurgy.

Abstracts of British Standards are included by permission of the British Standards Institution, 2 Park Street, London, from whom copies of the latest complete standards may be obtained.

The terms TIG, MIG and CO_2 have been retained for these welding processes, pending revision of BS 499, as they are so widely used. The use of gas mixtures of inert and active gases (argon–oxygen, argon–oxygen–CO_2, argon–hydrogen, etc.) as the shielding gas together with the pulsed, modulated feed, modulated arc length and flux cored processes, etc. have made the present terminology rather inadequate.

Oswestry
1977   A. C. Davies

The metric system and the use of SI units

The metric system was first used in France after the French Revolution and has since been adopted for general measurements by all countries of the world except the United States. For scientific measurements it is generally used universally.

It is a decimal system, based on multiples of ten, the following multiples and sub-multiples being added, as required, as a prefix to the basic unit.

Prefixes for SI units

Prefix	Symbol	Factor
atto	a	10^{-18}
femto	f	10^{-15}
pico	p	10^{-12}
nano	n	10^{-9}
micro	μ	10^{-6}
milli	m	10^{-3}
centi	c	10^{-2}
deci	d	10^{-1}
deca	da	10^{1}
hecto	h	10^{2}
kilo	k	10^{3}
mega	M	10^{6}
giga	G	10^{9}
tera	T	10^{12}
peta	P	10^{15}
exa	E	10^{18}

Examples of the use of these multiples of the basic unit are: hectobar, milliampere, meganewton, kilowatt.

In past years, the CGS system, using the centimetre, gram and second as the basic units, has been used for scientific measurements. It was later modified to the MKS system, with the metre, kilogram and second as the basic units, giving many advantages, for example in the field of electrical technology.

Note on the use of indices

A velocity measured in metres per second may be written m/s, indicating that the second is the denominator, thus: $\dfrac{\text{metre}}{\text{second}}$ or $\dfrac{\text{m}}{\text{s}}$. Since $\dfrac{1}{a^n} = a^{-n}$, the velocity can also be expressed as metre second^{-1} or m s^{-1}.

This method of expression is often used in scientific and engineering articles. Other examples are; pressure and stress: newton per square metre or pascal (N/m^2 or Nm^{-2}); density: kilograms per cubic metre (kg/m^3 or kg m^{-3}).

SI units (Système Internationale d'Unités)

To rationalize and simplify the metric system the Système Internationale d'Unités was adopted by the ISO (International Organization for Standardization). In this system there are six primary units, thus:

Quantity	Basic SI unit	Symbol
length	metre	m
mass	kilogram	kg
time	second	s
electric current	ampere	A
temperature	kelvin	K
luminous intensity	candela	cd

In addition there are derived and supplementary units, thus:

Quantity	Unit	Symbol
plane angle	radian	rad
area	square metre	m^2
volume*	cubic metre	m^3
velocity	metre per second	m/s
angular velocity	radian per second	rad/s
acceleration	metre per second squared	m/s^2
frequency	hertz	Hz
density	kilogram per cubic metre	kg/m^3
force	newton	N
moment of force	newton per metre	N/m
pressure, stress	newton per square metre	N/m^2 (of pascal, Pa)

Quantity	Unit	Symbol
surface tension	newton per metre	N/m
work, energy, quantity of heat	joule	J (N/m)
power, rate of heat flow	watt	w (J/s)
impact strength	joule per square metre	J/m^2
temperature	degree Celsius	°C
thermal coefficient of linear expansion	reciprocal degree Celsius or kelvin	$°C^{-1}$, K^{-1}
thermal conductivity	watt per metre degree C	W/m °C
coefficient of heat transfer	watt per square metre degree C	W/m^2 °C
heat capacity	joule per degree C	J/°C
specific heat capacity	joule per kilogram degree C	J/kg °C
specific latent heat	joule per kilogram	J/kg
quantity of electricity	coulomb	C (As)
electric tension, potential difference, electromotive force	volt	V (W/A)
electric resistance	ohm	Ω (V/A)
electric capacitance	farad	F
magnetic flux	weber	Wb
inductance	henry	H
magnetic flux density	tesla	$T(Wb/m^2)$
magnetic field strength	ampere per metre	A/m
magnetomotive force	ampere	A
luminous flux	lumen	lm
luminance	candela per square metre	cd/m^2
illumination	lux	lx

* *Note.* Nm^3 is the same as m^3 at normal temperature and pressure, i.e. 0°C and 760mm Hg (NTP or STP).

The litre is used instead of the cubic decimetre (1 litre $= 1 \, dm^3$) and is used in the welding industry to express the volume of a gas.

Prɘssure and stress may also be expressed in bar (b) or hectobar (hbar) instead of newton per square metre.

Conversion factors from British units to SI units are given in the appendix.

1 metric tonne $= 1000$ kg.

1

Welding science

Heat

Solids, liquids and gases: atomic structure

Substances such as copper, iron, oxygen and argon which cannot be broken down into any simpler substances are called elements; there are at the present time over 100 known elements. A substance which can be broken down into two or more elements is known as a compound.

An *atom* is the smallest particle of an element which can take part in a chemical reaction. It consists of a number of negatively charged particles termed electrons surrounding a massive positively charged centre termed the nucleus. Since like electric charges repel and unlike charges attract, the electrons experience an attraction due to the positive charge on the nucleus. Chemical compounds are composed of atoms, the nature of the compound depending upon the number, nature and arrangement of the atoms.

A molecule is the smallest part of a substance which can exist in the free state and yet exhibit all the properties of the substance. Molecules of elements such as copper, iron and aluminium contain only one atom and are monatomic. Molecules of oxygen, nitrogen and hydrogen contain two atoms and are diatomic. A molecule of a compound such as carbon dioxide contains three atoms and complicated compounds contain many atoms.

An atom is made up of three elementary particles: (1) protons, (2) electrons, (3) neutrons.

The *proton* is a positively charged particle and its charge is equal and opposite to the charge on an electron. It is a constitutent of the nucleus of all atoms and the simplest nucleus is that of the hydrogen atom, which contains one proton.

The *electron* is 1/1836 of the mass of a proton and has a negative charge equal and opposite to the charge on the proton. The electrons form a cloud around the nucleus moving within the electric field of the positive charge and around which they are arranged in shells.

The *neutron* is a particle which carries no electric charge but has a mass

1

equal to that of the proton and is a constituent of the nuclei of all atoms except hydrogen. The atomic number of an element indicates the number of protons in its nucleus and because an atom in its normal state exhibits no external charge, it is the same as the number of electrons in the shells.

Isotopes are forms of an element which differ in their atomic mass but not in some of their chemical properties. The atomic weight of an isotope is known as its mass number. For example, an atom of carbon has 6 protons and 6 neutrons in its nucleus so that its atomic number is 6. Other carbon atoms exist, however, which have 7 neutrons and 8 neutrons in the nucleus. These are termed isotopes and their mass numbers are 13 and 14 respectively, compared with 12 for the normal carbon atom. One isotope of hydrogen, called heavy hydrogen or deuterium, has a mass number 2 so that it has one proton and one neutron in its nucleus.

Electron shells. The classical laws of mechanics as expounded by Newton do not apply to the extremely minute world of the atom and the density, energy and position of the electrons in the shells are evaluated by quantum or wave mechanics. Since an atom in its normal state is electrically neutral, if it loses one or more electrons it is left positively charged and is known as a *positive ion*; if the atom gains one or more electrons it becomes a *negative ion*. It is the electrons which are displaced from their shells, the nucleus is unaffected, and if the electrons drift from shell to shell in an organized way in a completed circuit this constitutes an electric current.

In the *periodic classification*, the elements are arranged in order of their mass numbers, horizontal rows ending in the inert gases and vertical columns having families of related elements.

The lightest element, hydrogen, has one electron in an inner shell and the following element in the table, helium, has two electrons in the inner shell. This shell is now complete so that for lithium, which has three electrons, two occupy the inner shell and one is in the next outer shell. With succeeding elements this shell is filled with electrons until it is complete with the inert gas neon, which has two electrons in the inner shell and eight in the outer shell, ten electrons in all. Sodium has eleven electrons, two in the inner, eight in the second and one in a further outer shell. Electrons now fill this shell with succeeding elements until with argon it is temporarily filled with eight electrons so that argon has eighteen electrons in all. This is illustrated in Fig. 1.1 and this brief study will suffice to indicate how atoms of the elements differ from each other. Succeeding elements in the table have increasing numbers of electrons which fill more shells until the table is, at the present time, complete with just over 100 elements.

hydrogen	helium						
1	2						
lithium	beryllium	boron	carbon	nitrogen	oxygen	fluorine	neon
2	2	2	2	2	2	2	2
1	2	3	4	5	6	7	8
sodium	magnesium	aluminium	silicon	phosphorus	sulphur	chlorine	argon
2	2	2	2	2	2	2	2
8	8	8	8	8	8	8	8
1	2	3	4	5	6	7	8

The shells are then filled up thus:

2	8.				
2	8	8.			
2	8	18.			
2	8	18	8.		
2	8	18	18.		
2	8	18	18	8.	
2	8	18	32	18.	
2	8	18	32	18	8.

The electrons in their shells possess a level of energy and with any change in this energy light is given out or absorbed. The elements with completed or temporarily completed shells are the active or inert gases helium; neon, argon, xenon and radon, whereas when a shell is nearly complete (oxygen, fluorine) or has only one or two electrons in a shell (sodium, magnesium), the element is very reactive, so that the characteristics of an element are greatly influenced by its electron structure. When a metal filament such as tungsten is heated in a vacuum it emits electrons, and if a positively charged plate (anode) with an aperture in it is put in front near the filament, the electrons stream through the aperture attracted by the positive charge and form an electron beam. This beam can be focused and guided and is used in the television tube, while a beam of higher energy can be used for welding by the electron beam process (see Volume 2).

Fig. 1.1

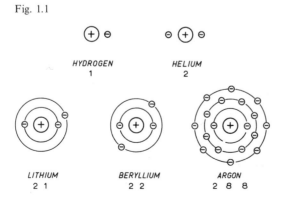

HYDROGEN
1

HELIUM
2

LITHIUM
2 1

BERYLLIUM
2 2

ARGON
2 8 8

If the atoms in a substance are not grouped in any definite pattern the substance is said to be amorphous, while if the pattern is definite the substance is crystalline. Solids owe their rigidity to the fact that the atoms are closely packed in geometrical patterns called space lattices which, in metals, are usually a simple pattern such as a cube. The positions which atoms occupy to make up a lattice can be observed by X-rays.

Atoms vibrate about their mean position in the lattice, and when a solid is heated the heat energy supplied increases the energy of vibration of the atoms until their mutual attraction can no longer hold them in position in the lattice so that the lattice collapses, the solid melts and turns into a liquid which is amorphous. If we continue heating the liquid, the energy of the atoms increases until those having the greatest energy and thus velocity, and lying near the surface, escape from the attraction of neighbouring atoms and become a vapour or gas. Eventually when the vapour pressure of the liquid equals atmospheric pressure (or the pressure above the liquid) the atoms escape wholesale throughout the mass of the liquid which changes into a gaseous state and the liquid boils.

Suppose we now enclose the gas in a closed vessel and continue heating. The atoms are receiving more energy and their velocity continues to increase so that they will bombard the walls of the vessel, causing the pressure in the vessel to increase.

Atoms are grouped into molecules, which may be defined as the smallest particles which can exist freely and yet exhibit the chemical properties of the original substance. If an atom of sulphur, two atoms of hydrogen, and four atoms of oxygen combine, they form a molecule of sulphuric acid. The molecule is the smallest particle of the acid which can exist, since if we split it up we are back to the original atoms which combined to form it.

From the foregoing, it can be seen that the three states of matter – solids, liquids and gases – are very closely related, and that by giving or taking away heat we can change from one state to the other. Ice, water and steam give an everyday example of this change of state.

Metals require considerable heat to liquefy or melt them, as for example, the large furnaces necessary to melt iron and steel.

We see examples of metals in the gaseous state when certain metals are heated in the flame. The flame becomes coloured by the gas of the metal, giving it a characteristic colour, and this colour indicates what metal is being heated. For example, sodium gives a yellow coloration and copper a green coloration.

This change of state is of great importance to the welder, since he is concerned with the joining together of metals in the liquid state (termed fusion welding) and he has to supply the heat to cause the solid metal to be converted into the liquid state to obtain correct fusion.

Temperature: thermometers and pyrometers

The temperature of a body determines whether it will give heat to, or receive heat from, its surroundings.

Our sense of determining hotness by touch is extremely inaccurate, since iron will always feel colder than wood, for example, even when actually at the same temperature.

Instruments to measure temperature are termed thermometers and pyrometers. Thermometers measure comparatively low temperatures, while pyrometers are used for measuring the high temperatures as, for example, in the melting of metals.

In the thermometer, use is made of the fact that some liquids expand by a great amount when heated. Mercury and alcohol are the usual liquids used. Mercury boils at 375°C and thus can be used for measuring temperatures up to about 330°C.

Mercury is contained in a glass bulb which connects into a very fine bore glass tube called a capillary tube and up which the liquid expands (Fig. 1.2).

The whole is exhausted of air and sealed off. The fixed points on a thermometer are taken as the melting point of ice and the steam from pure water at boiling point at standard pressure (760 mm mercury).

In the Celsius or Centigrade thermometer the freezing point is marked 0 and boiling point 100; thus there are 100 divisions, called degrees and shown thus °. The Kelvin scale (K) has its zero at the absolute zero of temperature, which is − 273.16°C. To convert approximately from °C to K add 273 to the Celsius figure.

Fig. 1.2. Celsius or Centigrade graduations.

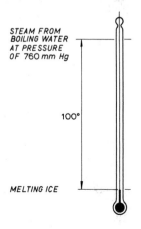

STEAM FROM
BOILING WATER
AT PRESSURE
OF 760 mm Hg

100°

MELTING ICE

To measure temperatures higher than those measurable with an ordinary thermometer we can employ:

(1) Temperature cones.
(2) Temperature-indicating paints or crayons.
(3) Pyrometers:
    (*a*) Electrical resistance.
    (*b*) Thermo-electric.
    (*c*) Radiation.
    (*d*) Optical.

(1) Temperature cones (Seger cones) are triangular pyramids made of a mixture of china clay, lime, quartz, iron oxide, magnesia, and boric acid in varying proportions so that they melt at different temperatures and can be used to measure temperatures between 600 °C and 2000 °C. They are numbered according to their melting points and are generally used in threes, numbered consecutively, of approximately the temperature required. When the temperature reaches that of the lowest melting point cone it bends over until its apex touches the floor. The next cone bends slightly out of the vertical while the third cone remains unaffected. The temperature of the furnace is that of the cone which has melted over.

(2) Temperature-indicating paints and crayons either melt or change colour or appearance at definite temperatures. Temperature indicators are available as crayons (sticks), pellets or in liquid form and operate on the melting principle and not colour change. They are available in a range from 30 °C to 1650 °C and each crayon has a calibrated melting point. To use the crayon, one of the temperature range required is stroked on the work as the temperature rises and leaves a dry opaque mark until at the calibrated temperature it leaves a liquid smear which on cooling solidifies to a translucent or transparent appearance. Up to 700 °C a mark can be made on the work piece before heating and liquefies at the temperature of the stick. Similarly a pellet of the required temperature is placed on the work and melts at the appropriate temperature while the liquid is sprayed on to the surface such as polished metal (or glass) which is difficult to mark with a crayon, and dries to a dull opaque appearance. It liquefies sharply at its calibrated temperature and remains glossy and transparent upon cooling.

(3) Pyrometers. (*a*) Electrical resistance pyrometers. Pure metals increase in resistance fairly uniformly as the temperature increases. A platinum wire is wound on a mica former and is placed in a refractory sheath, and the unit placed in the furnace. The resistance of the platinum wire is measured (in a Wheatstone's bridge network) by passing a current through it. As the temperature of the furnace increases the resistance of the platinum

increases and this increase is measured and the temperature read from a chart.

(*b*) Thermo-electric (thermo-couple) pyrometers. When two dissimilar metals are connected together at each end and one pair (or junction) of ends is heated while the other pair is kept cold, an electromotive force (e.m.f.) or voltage is set up in the circuit (Peltier Effect). The magnitude of this e.m.f. depends upon (*a*) the metals used and (*b*) the difference in temperature between the hot and cold junctions. In practice the hot junction is placed in a refractory sheath while the other ends (the cold junction) are connected usually by means of compensating leads to a millivoltmeter which measures the e.m.f. produced in the circuit and which is calibrated to read the temperature directly on its scale. The temperature of the cold junction must be kept steady and since this is difficult, compensating leads are used. These are made of wires having the same thermo-electric characteristics as those of the thermo-couple but are much cheaper and they get rid of the thermo-electric effect of the junction between the thermo-couple wires and the leads to the millivoltmeter, when the temperature of the cold junction varies. The couples generally used are copper-constantan (60% Cu, 40% Ni) used up to 300 °C; chromel (90% Ni, 10% Cr); alumel (95% Ni, 3% Mn, 2% Al) up to 1200 °C; and platinum–platinum–rhodium (10% Rh) up to 1500 °C (Fig. 1.3*a*).

Fig. 1.3

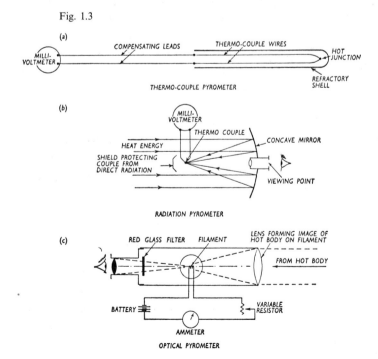

(*c*) Radiation pyrometers. These pyrometers measure the radiation emitted from a hot body. A 'black body' surface is one that absorbs all radiation falling upon it and reflects none, and conversely will emit all radiations. For a body of this kind, E, the heat energy radiated, is proportional to the fourth power of the absolute temperature, i.e. $E \propto T^4$ (Stefan–Boltzmann Law) so that $E = kT^4$. If a body is however radiating heat in the open, the ratio of the heat which it radiates to the heat that a black body would radiate at the same temperature is termed the emissivity, *e*, and this varies with the nature, colour and temperature of the body. Knowing the emissivity of a substance we can calculate the true temperature of it when radiating heat in the open from the equation:

$$(\text{True temperature})^4 = \frac{(\text{Apparent temperature})^4}{\text{emissivity}}$$

(temperatures are on the absolute scale).

In an actual radiation pyrometer the radiated heat from the hot source is focussed on to a thermo-couple by means of a mirror (the focussing can be either fixed or adjustable) and the image of the hot body must cover the whole of the thermo-couple. The e.m.f. generated in the thermo-couple circuit is measured as previously described on a millivoltmeter (Fig. 1.3*b*).

(*d*) Optical pyrometers. The disappearing filament type is an example of this class of pyrometer. A filament contained in an evacuated bulb like an electric light bulb is viewed against the hot body as a background. By means of a control resistor the colour of the filament can be varied by varying the current passing through it until the filament can no longer be seen, hot body and filament then being at the same temperature. An ammeter measures the current taken by the filament and can be calibrated to read the temperature of the filament directly.

The judging of temperatures by colour is usually very inaccurate. If steel is heated, it undergoes a colour change varying from dull red to brilliant white. After considerable experience it is possible to estimate roughly the temperature by this means, but no reliance can be placed on it (Fig. 1.3*c*).

Temperature gradient and heat affected zone

The gradient is the rate of change of a quantity with distance so that the temperature gradient along a metal bar is the rate of change of temperature along the bar.

This can be illustrated by Fig. 1.4*a*, which shows the graph of temperature plotted against distance for a bar heated at one end, the other end being cold. The hot portion loses heat by conduction, convection and radiation. The conduction of heat along the bar will be greater the better the thermal conductivity of the metal of the bar so that heat travels more quickly along a copper bar than along a steel bar, and the graph shows that the hot end is losing heat to its surroundings much more quickly than the colder parts of the bar. The greater the difference of temperature between two points, the more rapidly is the heat lost.

If two steel plates are welded together, Fig. 1.4*b* shows the temperature gradient from the molten pool to the cold parent plate on each side of the weld. The gradient on one side only is shown in the figure. That portion of

Fig. 1.4

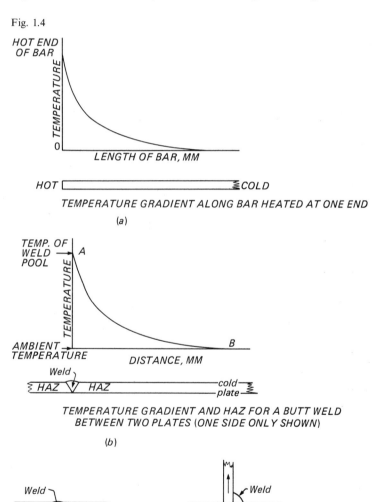

TEMPERATURE GRADIENT ALONG BAR HEATED AT ONE END

(a)

TEMPERATURE GRADIENT AND HAZ FOR A BUTT WELD BETWEEN TWO PLATES (ONE SIDE ONLY SHOWN)

(b)

HEAT DISSIPATION BY CONDUCTION

(c)

the plate on either side of the weld, affected by the heat and in which the metal suffers thermal disturbance, is known as the Heat-Affected Zone (HAZ) and the areas nearest the weld in which this disturbance is greatest undergo a change in structure which may include recrystallization, refining and grain growth (q.v.). The larger and thicker the plate the more quickly will the molten pool lose heat (or freeze) by conduction and for this reason an arc should never be struck briefly on a thick section cold plate especially in certain steels as the sudden quenching effect may lead to cracking.

Fig. 1.4c shows that a butt weld loses heat by conduction in two directions while in a fillet joint the heat has three directions of travel. Both joints also lose heat by convection and radiation.

Expansion and contraction

When a solid is heated, the atoms of which it is composed vibrate about their mean position in the lattice more and more. This causes them to take up more room and thus the solid expands.

Most substances expand when heated and contract again when cooled, as the atoms settle back into their normal state of vibration.

Metals expand by a much greater amount than other solid substances, and there are many practical examples of this expansion in everyday life.

Gaps are left between lengths of railway lines, since they expand and contract with atmospheric temperature changes. Fig. 1.5a shows the expansion joint used by British Rail. With modern methods of track construction only the last 100 m of rail is allowed to expand or contract longitudinally irrespective of the total continuous length of welded rail, and this movement is well within the capacity of the expansion joint or adjustment switch.

Iron tyres are made smaller than the wheel they are to fit. They are heated and expand to the size of the wheel and are fitted when hot. On being quickly cooled, they contract and grip the wheel firmly.

Large bridges are mounted on rollers fitted on the supporting pillars to allow the bridge to expand.

In welding, this expansion and contraction is of the greatest importance. Suppose we have two pieces of steel bar about 1 m long. If these are set together at an angle of 90 °, as shown, and then welded and allowed to cool, we find that they have curled or bent up in the direction of the weld (Fig. 1.5b and c).

The hot weld metal, on contracting, has caused the bar to bend up as shown, and it is evident that considerable force has been exerted to do this.

A well-known example of the use to which these forces, exerted during expansion and contraction, are put is the use of iron bars to pull in or strengthen defective walls of buildings.

Plates or S pieces are placed on the threaded ends of the bar, which projects through the walls which need pulling in. The bar is heated to redness and nuts on each end are drawn up tight against the plates on the walls. As the bar cools, gradually the walls are pulled in.

Different metals expand by different amounts. This may be shown by riveting together a bar of copper and a bar of iron about 0.5 m long and 25 mm wide. If this straight composite bar is heated it will become bent, with the copper on the outside of the bend, showing that the copper expands more than the iron (Fig. 1.6). This composite bar is known as a bi-metal strip and is used in engineering for automatic control of temperature.

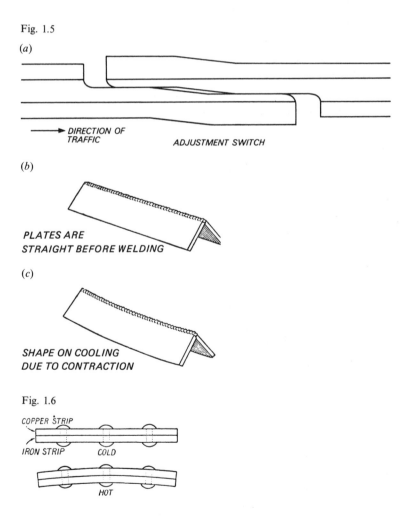

Fig. 1.5

(a)

DIRECTION OF TRAFFIC

ADJUSTMENT SWITCH

(b)

PLATES ARE STRAIGHT BEFORE WELDING

(c)

SHAPE ON COOLING DUE TO CONTRACTION

Fig. 1.6

COPPER STRIP

IRON STRIP   COLD

HOT

Coefficient of linear expansion

The fraction of its length which a bar will expand when heated through one degree rise in temperature is termed its *coefficient of linear expansion*. (This also applies to contraction when the bar is cooled.) This fraction is very small; for example, for iron it is

$$\frac{12}{1\,000\,000}$$

That is, a bar of iron length l would expand by $l \times \dfrac{12}{1\,000\,000}$ for every degree rise in temperature. Hence, if the rise was $t°$, the expansion would be $\dfrac{12}{1\,000\,000} \times l \times t$.

The fraction $\dfrac{12}{1\,000\,000}$ is usually denoted by the letter a. Thus the increase in length of a bar of original length l, made of material whose coefficient of linear expansion is a, when heated through $t°$ is lat.

Thus, the final length of a bar when heated equals its original length plus its expansion, that is:

$$L = l + lat$$
Final length = original length + expansion.

This can also be written: $L = l\,(1 + at)$.

Length after being
heated through $t°C$

Example

Given that the coefficient of linear expansion of copper is $\dfrac{17}{1\,000\,000}$ or $0.000\,017$ per degree C, find the final length of a bar of copper whose original length was 75 mm, when heated through 50°C.

Final length = original length + expansion, i.e.:

$$\text{Final length} = 75 + \left(75 \times \frac{17}{1\,000\,000} \times 50\right)$$

$$= 75 + \frac{63\,750}{1\,000\,000} = 75 + \frac{6375}{100\,000} = 75.06 \text{ mm.}$$

The above is equally true for calculating the contraction of a bar when cooled.

Table of coefficients of linear expansion of metals per degree C

Metal	a	Metal	a
Lead	0.000 027	Zinc	0.000 026
Tin	0.000 021	Cast iron	0.000 010
Aluminium	0.000 025	Nickel	0.000 013
Copper	0.000 017	Wrought iron	0.000 012
Brass	0.000 020	Mild steel	0.000 012
60% copper, 40% zinc			

Invar, a nickel–steel alloy containing 36% nickel, has a coefficient of linear expansion of only 0.000 000 9, that is, only $\frac{1}{13}$ of that of mild steel, and thus we can say that invar has practically no expansion when heated.

The expansion and contraction of metal is of great importance to the welder, because, as we have previously shown, large forces or stresses are called into play when it takes place. If the metal that is being welded is fairly elastic, it will stretch, or give, to these forces, and this is a great help, although stresses may be set up as a result in the welded metal. Some metals, however, like cast iron, are very brittle and will snap rather than give or show any elasticity when any force is applied. As a result, the greatest care has to be taken in applying heat to cast iron and in welding it lest we introduce into the metal, when expanding and contracting, any forces which will cause it to break. This will be again discussed at a later stage.

Coefficient of cubical expansion

If we imagine a solid being heated, it is evident that its volume will increase, because each side undergoes linear expansion.

A cube, for example, has three dimensions, and each will expand according to the previous rule for linear expansion. Suppose each face of the cube was originally length l and finally length L after being heated through $t°$C. Let the coefficient of linear expansion be a per degree C.

The original volume was $l \times l \times l = l^3$.

Each edge will have expanded, and for each edge we have:

Final length $L = l(1 + at)$ as before (Fig. 1.7).

Thus the new volume $= l(1 + at) \times l(1 + at) \times l(1 + at)$
$$= l^3(1 + 3at) \text{ approximately.}$$

Thus, the final volume = original volume $(1 + 3at)$.

That is, the *coefficient of cubical expansion* may be taken as being three times the coefficient of linear expansion.

Example

A brass cube has a volume of 0.006 m³ (6×10^6 mm³) and is heated through a 65 °C rise in temperature. Find its final volume, given that the coefficient of linear expansion of brass = 0.000 02 per degree C.

Final volume = original volume $(1 + 3at)$
$$V = 0.006(1 + 3 \times 0.000\,02 \times 65)$$
$$= 0.006(1.0039)$$
$$= 0.006\,023\ 4\text{m}^3.$$

The joule and the newton

Heat is a form of energy and the unit of energy is the joule (J). A joule may be defined as the energy expended when a force of 1 newton (N) moves through a distance of 1 metre (m). (Note: a newton is that force which, acting on a mass of 1 kilogram (kg), gives it an acceleration of 1 metre per second per second (1 m/s²). The gravitational force on a mass of 1 kg equals 9.81 N so that for practical purposes, to convert from kilograms force to newtons, multiply by 10.)

Specific heat capacity

The specific heat capacity is defined as the quantity of heat required to raise unit mass of a substance through 1° rise in temperature. The SI unit is in kilojoules per kilogram K (kJ/kg K, symbol c). Note that since K adds 273 to °C, a rise in temperature will be the same in K or °C.

Example

Find the heat gained by a mass of 20 kg of cast iron which is raised through a temperature of 30 °C, given that the specific heat capacity of cast iron is 0.55 kJ/kg K.

Fig. 1.7

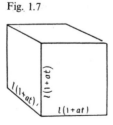

Heat gained = mass × c × rise in temperature
= 20 × 0.55 × 30 kilojoules
= 330 kJ.

Substance	Specific heat capacity	Substance	Specific heat capacity
Water	4.2×10^3	Mild steel	0.45×10^3
Aluminium	0.91×10^3	Wrought iron	0.47×10^3
Tin	0.24×10^3	Zinc	0.4×10^3
Lead	0.13×10^3	Cast iron	0.55×10^3
Copper	0.39×10^3	Nickel	0.46×10^3
Brass	0.38×10^3		

The above values are approximately 4.2×10^3 as great as the values when specific heat capacities were expressed in calories per gram degree C.

Melting point

The melting point of a substance is the temperature at which the change of state from solid to liquid occurs, and this is usually the same temperature at which the liquid will change back to solid form or freeze.

Substances which expand on solidifying have their freezing point lowered by increase of pressure while others which contract on freezing have their freezing point raised by pressure increase.

The melting point of a solid with a fairly low melting point can be determined by attaching a small glass tube, with open end containing some of the solid, to the bulb of a thermometer. The thermometer is then placed in a container holding a liquid, whose boiling point is above the melting point of the solid, and fitted with a cover, as shown in Fig. 1.8, and a stirrer is also included. The container is heated and the temperature at which the solid melts is observed. The apparatus is now allowed to cool and the temperature at which the substance solidifies is noted. The mean of these two readings gives the melting point of the solid. By using mercury, which boils at 357 °C, as the liquid in the container, the melting point of solids which melt between 100 and 300 °C could be obtained.

Determination of the melting point by method of cooling

The solid, of which the melting point is required, is placed in a suitable container, fitted with a cork or stopper through which a thermometer is inserted (Fig. 1.9). A hole in the stopper prevents pressure rise. The container is heated until the solid melts, and heating is continued until the temperature is raised well above this point. The liquid is now allowed to cool and solidify and the temperature is taken every quarter or

Fig. 1.8

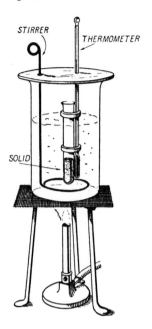

Fig. 1.9

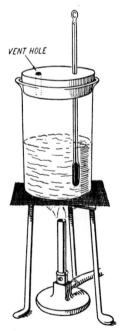

half minute. This temperature is plotted on a graph against the time, and the shape of the graph should be as shown in Fig. 1.10.

If the melting point of a metal is required, the metal is placed in a fireclay or graphite crucible and heated by means of a furnace, and the temperature is measured, at the same intervals, by a pyrometer. The metal, on cooling, begins to solidify and form crystals in exactly the same way as any other solid. The portion *A* shows the fall in temperature of the liquid or molten metal. The portion *B* indicates the steady temperature while solidification is taking place, and portion *C* shows the further fall in temperature as the solid loses heat. The temperature $t°$ of the portion *B* of the curve is the melting point of the solid.

In practice we may find that the temperature falls below the dotted line, as shown, that is, below the solidifying temperature. This is due to the difficulty which the liquid may experience in commencing to form crystals, and is called 'super-cooling'. It then rises again to the true solidifying point and cooling then takes place as before (Fig. 1.11).

This method of determination of the melting point is much used in finding the melting point of alloys and in observing the behaviour of the constituents of the alloys when melting and solidifying.

The melting point of a metal is of great importance in welding, since,

Fig. 1.10

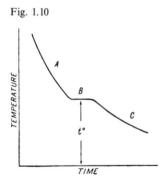

Fig. 1.11

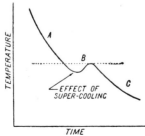

together with the capacity for heat of the metal, it determines how much heat is necessary for fusion. The addition of other substances or metals to give a metal (thus forming an alloy) will affect its melting point.

Specific latent heat

If a block of ice is placed in a vessel with a thermometer and heat is applied, the temperature remains steady at $0\,°C$ ($273\ K$) until the whole of the ice has been melted and then the temperature begins to rise. The heat given to the ice has not caused any rise in temperature but a change of state from solid to liquid and is called the specific latent heat of fusion. When the change of state is from liquid to gas it is termed the specific latent heat of vaporization (or evaporation) and is expressed in joules or kilojoules (kJ) per kilogram (J/kg or kJ/kg).

Specific latent heat of fusion in kJ/kg

Aluminium	393	Nickel	273
Copper	180	Tin	58
Iron	205	Ice	333

Specific latent heat of fusion is more important in welding than specific latent heat of vaporization, because a comparison of these figures gives an indication of the relative amounts of heat required to change the solid metal into the liquid state before fusion.

Since the heat must be given to a solid to convert it to a liquid, it follows that heat will be given out by a liquid when solidifying. This has already been demonstrated when determining the melting point of a liquid by the method of cooling. When the change of state from liquid to solid takes place (*B* on the curve in Fig. 1.10) heat is given out and the temperature remains steady until solidification is complete, when it again begins to fall.

Transfer of heat

Heat can be transferred in three ways: conduction, convection, radiation.

Conduction. If the end of a short piece of metal rod is heated in a flame, it rapidly gets too hot to hold (Fig. 1.12). Heat has been transferred by conduction from atom to atom through the metal from the flame to the hand. If a rod of copper and one of steel are placed in the flame, the copper rod gets hotter more quickly than the steel one, showing that the heat has been conducted by the copper more quickly than the steel. If the rods are held in a cork and the cork gripped in the hand, they can now be held comfortably. The cork is a bad conductor of heat. All metals are good conductors but some are better than others, and the rate at which heat is conducted is termed the *thermal conductivity* and is measured in watts per metre degree (W/m °C).

The conductivity depends on the purity of the metal, its structure and the temperature.

As the temperature rises the conductivity decreases and impurities in a metal greatly reduce the conductivity.

The thermal conductivity is closely allied to the electrical conductivity, that is, the ease with which an electric current is carried by a metal. It is interesting to compare the second and third columns in the table. From these we see that in general the better a metal conducts heat, the better it conducts electricity.

Table of comparative conductivities (taking copper as 100)

	Thermal conductivity	Electrical conductivity
Silver	106	108
Copper	100	100
Aluminium	62	56
Zinc	29	29
Nickel	25	15
Iron	17	17
Steel	13–17	13–17
Tin	15	17
Lead	8	9

Fig. 1.12

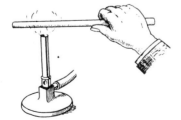

The effect of conductivity of heat on welding practice can clearly be seen from the calculations in Fig. 1.13, where a block of copper and one of steel of equal mass are to be welded. It is seen that if the two blocks were to be each brought up throughout their mass to melting point, the steel would take *a much greater quantity of heat* than the copper would.

When the heat is applied at one spot, copper being such a good conductor, heat is rapidly transferred from this spot throughout its mass, and we find that the spot where the heat is applied will not melt until the whole mass of the copper has been raised to a very high temperature indeed.

With the mild steel block, on the other hand, the heat conductivity is only about $\frac{1}{6}$ (from the table) that of the copper, that is, the heat is conducted away at only $\frac{1}{6}$ the rate. Hence we find that the spot where the heat is applied will be raised to melting point long before the rest of the block has become very hot.

Because of this high conductivity of copper, it is usual to employ greater heat than when welding the same thickness of steel or iron.

For this reason also, when welding copper, whether by arc or oxy-acetylene, it is always advisable to heat the work up to a high temperature over a large area around the area to be welded. In this way the heat will not be conducted to colder regions so rapidly and better fusion in the weld itself can be obtained.

Cast iron is a comparatively poor conductor of heat compared with

Fig. 1.13

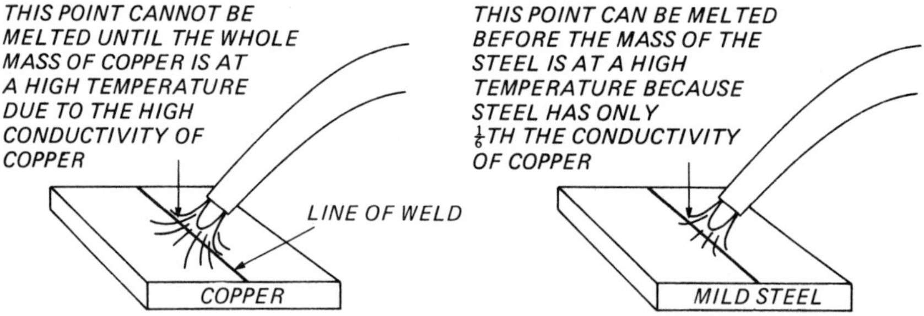

THIS POINT CANNOT BE MELTED UNTIL THE WHOLE MASS OF COPPER IS AT A HIGH TEMPERATURE DUE TO THE HIGH CONDUCTIVITY OF COPPER

LINE OF WELD

COPPER

THIS POINT CAN BE MELTED BEFORE THE MASS OF THE STEEL IS AT A HIGH TEMPERATURE BECAUSE STEEL HAS ONLY $\frac{1}{6}$TH THE CONDUCTIVITY OF COPPER

MILD STEEL

Mass 1 kg
Melting point of copper 1083 °C
Specific heat capacity 0.39 ×
  10^3 J/kg °C
Heat required to raise copper
  to melting point
  = $1 \times 1083 \times 0.39 \times 10^3$ J
  = 422 370 J
  = 0.422×10^6 J.

Mass 1 kg
Melting point of steel 1400 °C
Specific heat capacity 0.45 ×
  10^3 J/kg °C
Heat required to raise steel to
  melting point
  = $1 \times 1400 \times 0.45 \times 10^3$ J
  = 630 000 J
  = 0.63×10^6 J.

copper. If we heat a casting in one spot, therefore, heat will only be transferred away slowly. The part being heated thus expands more quickly than the surrounding parts and, since expansion is irregular, great forces, as before explained, are set up and, since cast iron is brittle and has very small elasticity, the casting fractures. The welding of cast iron is thus a study of expansion and contraction and conduction of heat and, to weld cast iron successfully, care must be taken that the temperature of the whole casting is raised and lowered equally throughout its mass. This will be discussed at a later stage.

Convection. When heat is transferred from one place to another by the motion of heated particles, this is termed convection. For example, in the hot water system of a house, heat from the fire heats the water and hot water, being less dense than colder water, rises in the pipes, forming convection currents and transferring heat to the storage tank.

In the heat treatment of steel it is often necessary to cool the steel slightly more quickly than if it cooled naturally, in order to harden it. It is cooled, therefore, in an air blast, the heat being transferred thus by convection.

Radiation. Heat is transferred by radiation as pulses of energy, termed quanta, through the intervening space. We sit in front of a fire and it feels warm. There is no physical contact between our bodies and the fire. The heat is being transferred by radiation. Heat transferred in this manner travels according to the laws of light and is reflected and bent in the same way.

The sun's heat is transmitted by radiation to our planet but the method by which the heat travels through space is not fully understood. Metal, if allowed to cool in a still atmosphere, loses its heat by radiation and any other bodies in the neighbourhood will become warmed.

It is evident that the outside of the hot metal will lose heat more quickly than the interior, and we find, for example, that the surface of cast iron is much harder than below the surface, because it has lost heat more quickly.

Chills are strips or blocks of metal placed adjacent to the line of weld during the welding operation in order to dissipate heat and reduce the area affected by the input of heat, the heat-affected zone (HAZ). Heat is removed by conduction, convection and radiation, and copper is often used because of its good heat conducting properties. Heat control can be effected by moving the chills nearer or further from the weld.

Behaviour of metals under loads

Stress, strain and elasticity

When a force, or load, is applied to a solid body it tends to alter the shape of the body, or deform it.

The atoms of the body, owing to their great attraction for each other, resist, up to a certain point, the attempt to alter their position and there is only a slight distortion of the crystal lattice.

If the applied force is removed before this point is reached, the body will regain its original shape.

This property, which most substances possess, of regaining their original shape upon removal of the applied load is termed *elasticity*.

Should the applied load be large enough, however, the resistance of the atoms will be overcome and they will move and take up new positions in the lattice. If the load is now removed, the body will no longer return to its former shape. It has become permanently distorted (Fig. 1.14).

The point at which a body ceases to be elastic and becomes permanently distorted or set is termed the yield point, and the load which is applied to cause this is the yield-point load. The body is then said to have undergone plastic deformation or flow.

Whenever a change of dimensions of a body occurs, from whatever cause, a state of *strain* is set up in that body. Strain is usually measured (for calculation purposes) by the ratio or fraction:

$$\frac{\text{change of dimensions in direction of applied load}}{\text{original dimensions in that direction}}$$

Fig. 1.14 (*a*) Original length of specimen. (*b*) Extension produced = l_1. Elastic limit not reached. F_1 = applied force. (*c*) Force removed, specimen recovers its original dimensions. (*d*) Extension produced = l_2. Elastic limit exceeded by application of force. Specimen now remains permanently distorted or set, and does not recover its original dimensions when force is removed. F_2 = applied force.

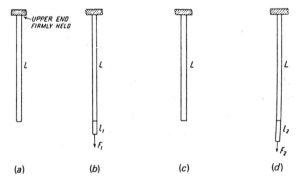

(a)   (b)   (c)   (d)

Example

A bar is 100 mm long and is stretched $\frac{1}{4}$ mm by an applied load along its length. Find the strain.

$$\text{Strain} = \frac{\text{change in length}}{\text{original length}} = \frac{\frac{1}{4}}{100} = \frac{1}{400}.$$

The magnitude of the force or load on unit area of cross-section of the body producing the strain is termed the *stress*.

Stress = force or load per unit area.

Hooke's Law states that for an elastic body strain is proportional to stress.

The mass of a body is the quantity of matter which it contains, so that it is dependent upon the number of atoms in its structure. Mass is measured in kilograms (kg) and 1000 grams (g) equal 1 kg. Note 1 lb = 0.4536 kg and 1 kg = 2.2 lb. Newton's Universal Law of Gravitation states that every particle of matter attracts every other particle of matter with a force (F) which is proportional to the product of the masses (m_1 and m_2) of the two particles and inversely proportional to the square of the distance (d) between them, $F \propto m_1 m_2 / d^2$. The weight of a body is the force by which it is attracted to the earth (the force of gravity), but because the earth is a flattened sphere, this force and hence the weight of the body vary somewhat according to its position on the earth's surface. On the surface of the moon, which has about one-sixth of the mass of the earth, a mass of one kilogram would weigh about one-sixth of a kilogram. To distinguish a mass of one kilogram from a force of one kilogram, which is the force of attraction due to the gravitational pull of the earth, the letter f is added thus, kgf.

The unit of force termed the newton avoids the distinction between mass and weight and is defined as 'that force which will give an acceleration of 1 metre per second per second to a mass of 1 kilogram'.

Units of stress or pressure

The following multiples of units are used:

tera- (T) = one million million	10^{12}
giga- (G) = one thousand million	10^9
mega- (M) = one million	10^6
kilo- (k) = one thousand	10^3
hecto- (h) = one hundred	10^2.

The SI unit of stress or pressure is the newton per square metre (N/m^2) which is also known as the pascal (Pa), and $1 \ N/m^2 = 1 \ Pa$. This is a small unit and when using it to express tensile strengths of materials large numbers are involved with the use of the meganewton per square metre (MN/m²) or megapascal (MPa).

If, however, the newton per square millimetre (N/mm²) is used, as in this book, large figures are avoided and the change to the SI unit is easily made since 1 N/mm² = 1 MN/m² or 1 MPa.

The bar (b) and its multiple the hectobar (hbar) are also used as units of pressure and stress. 1 bar is equal to the pressure of a vertical column of mercury 750 mm high and for conversion purposes it can be taken to equal 15 lbf* per square inch. It should be noted that 1 bar = 10⁵ N/m² or 10⁵ Pa, and 1 hbar = 10 N/mm².

Gauges for cylinders of compressed gases can be calibrated in bar, a cylinder pressure of 2500 lbf/in² being 172 bar.

Tensile strength can be expressed in hbar. A specimen of aluminium may have a tensile strength of 12 hbar which is equal to 120 N/mm².

If stress is stated in tonf/in² or kgf/mm² the following conversions can be used. (A full list of conversion factors is given in the appendix.)

Tonf/in² to MN/m² or N/mm², multiply by 15.5; MN/m² or N/mm² to tonf/in², multiply by 0.0647; kgf/m² to N/m², multiply by 9.8; and approximately 1 hbar = 1 kgf/mm².

If a stress is applied to a body and it changes its shape within its elastic limits, the ratio stress/strain is termed the modulus of elasticity or Young's modulus (E) of the material. The unit is N/m² or Pa, and a typical value for a specimen of aluminium is 69×10^3 MN/m² or MPa or N/mm².

There are three kinds of simple stress: (1) tensile, (2) compression, (3) shear.

Tensile stress

If one end of a metal rod is fixed firmly and a force is applied to the other end to pull the rod, it stretches. A tensile force has been applied to the rod and when it is measured on unit cross-sectional area it is termed a tensile stress.

Example

A force of 0.5 MN is applied so as to stretch a bar of cross-sectional area 400 mm². Find the tensile stress.

$$\text{Tensile stress} = \frac{\text{load}}{\text{area of cross-section}} = \frac{500\,000}{400} = 1250 \text{ N/mm}^2.$$

---

* 1 bar   = 14.508 lbf/in².
  1 lbf/in² = 0.0689 bar.

A machine known as a tensile strength testing machine, which will be described later (Chapter 5), is used for determining the tensile strength of materials and welded joints.

The specimen under test is clamped between two sets of jaws, one fixed and one moving, and the force can be increased until the specimen breaks.

Suppose a piece of mild steel is placed in the machine. As the tensile stress is increased, the bar becomes only very slightly longer for each increase of force. Then a point is reached when, for a very small increase of force, the bar becomes much longer. This is the yield point and the bar has been stretched beyond its elastic limit, and is now deforming plastically.

If the applied load had been reduced before this point was reached, the bar would have recovered its normal size, but will not do so when the yield point has been passed.

As the load is increased beyond the yield point the elongation of the bar for the same increase of loading becomes much greater, until a point is reached when the bar begins to get reduced in cross-sectional area and forms a waist, as shown. Less load is now required to extend the bar, since the load is now applied on a smaller area, the waist becomes smaller and the bar breaks. The accompanying diagram (Fig. 1.15) will make this clear.

When the stress is first applied, the extension of the bar is very small and needs accurate measurement, but it is proportional to the load so that the graph of stress/strain is a straight line. At one point X the graph deviates a little from the straight line OX, so that after X the strain is no longer proportional to the stress. The point X is the *limit of proportionality* and Hooke's Law is no longer obeyed. At Y the extension suddenly becomds

Fig. 1.15. Stress–strain diagram, mild steel.

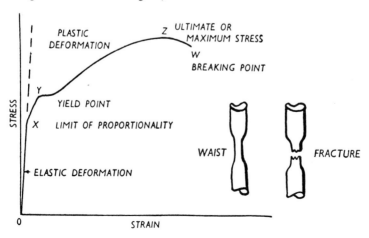

much greater than before for an equal increase in load and Y is termed the yield point, the stress at this point being the yield-point stress.

Increase of load produces progressive increase of length to the point Z. At this point the waist forms; Z is the maximum load. Breakage occurs at W under a smaller load than at Z. A substance which has a fair elongation during the plastic stage is called ductile, while if the elongation is very small it is said to be brittle.

Given a table of tensile strengths of various metals, we can calculate the maximum force or stress that any given section will stand.

Table of tensile strengths

The tensile strength of a metal depends upon its condition, whether cast, annealed, work-hardened, heat-treated, etc.

	N/mm²	hbar	tons f/in²		N/mm²	hbar	tonsf/in²
Lead	12–22	1.2–2.2	0.8–1.4	Brass	220–340	22–34	14–22
Zinc	30–45	3–4.5	2–3	Cast iron	220–300	22–30	14–20
Tin	30	3	2	Wrought iron	250–300	25–30	16–20
Aluminium	60–90	6–9	4–6	Mild steel	380–450	38–45	25–30
Copper	220–300	20–30	14–20	High tensile steel	600–800	60–80	

Example

A certain grade of steel has a tensile strength of 450 N/mm². What tensile force in newtons will be required to break a specimen of this steel cross-section 25 mm × 20 mm?

Area of cross-section = 500 mm²

Force required = tensile strength = area of cross-section
$$F = 450 \times 500 \text{ newtons}$$
$$= 225\,000 \text{ newtons}$$
$$= 225 \times 10^3 \text{ newtons} = 0.225 \text{ MN.}$$

The tensile strength of a metal depends largely upon the way it has been worked (hammered, rolled, drawn, etc.) during manufacture, its actual composition and the presence of impurities (see Fig. 1.16).

From the tensile test we can obtain:

(1) Yield point = $\dfrac{\text{yield stress}}{\text{original area of cross-section}}$ (N/mm² or hbar).

(2) Ultimate tensile stress (UTS)

$= \dfrac{\text{maximum stress}}{\text{original area of cross section}}$ (N/mm² or hbar).

(3) Percentage elongation on length between gauge marks

$$= \frac{\text{extension}}{\text{original length between gauge marks}} \times 100.$$

The distance between the gauge marks can be 50 mm or 5 × diameter of the specimen. Standard areas of cross-section can be 75mm² or 150 mm².

(4) Percentage reduction of area (R of A)

$$= \frac{\text{reduction of area at the fracture}}{\text{original area}} \times 100.$$

A typical example for one particular grade of weld metal is: Composition: 0.07% C, 0.4% Si, 0.68% Mn, remainder Fe. Yield stress 479 N/mm², ultimate tensile stress 556 N/mm², elongation on gauge length of 5 × *D*, 26%; reduction of area, 58%.

The elongation will depend on the gauge length. The shorter this is the greater the percentage elongation, since the greatest elongation occurs in the short length where 'waisting' or 'necking' has occurred. Reduction in area and elongation are an indication of the ductility of a metal.

As temperature rises there is usually a decrease in tensile strength and an increase in elongation, and the limit of proportionality is reduced so that at red heat application of stress produces plastic deformation. A fall in temperature usually produces the opposite effect. Internal stresses which

Fig. 1.16. Stress–strain diagram, for a steel in (1) annealed , (2) cold drawn condition.

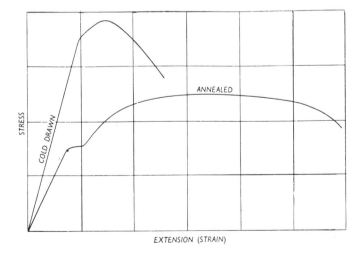

ANNEALED

STRESS

COLD DRAWN

EXTENSION (STRAIN)

have been left in a welded structure can be relieved by heating the members and lowering the limit of proportionality. The stresses then produce plastic deformation, and are relieved. This stress relief, however, may cause distortion.

Proof stress

Non-ferrous metals, such as aluminium and copper, etc., and also very hard steels, do not show a definite yield point, as just explained, and load–extension curves are shown in Fig. 1.17. A force which will produce a definite permanent extension of 0.1% or 0.2% of the gauge length is known as the proof stress (Fig. 1.18) and is measured in N/mm² or hbar.

Fig. 1.17

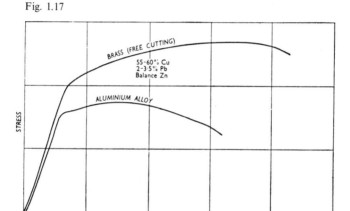

Fig. 1.18. Load–extension curve of hard steels and non-ferrous metals illustrating proof stress.

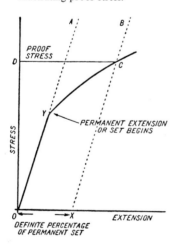

Compressive stress

If the forces applied in the previous experiments on tensile strength are reversed, the body is placed under compression.

Compressive tests are usually performed on specimens having a short length compared with their diameter to prevent buckling when the load is applied. Ductile metals increase in diameter to a barrel shape and cracking round the periphery is some indication of the ductility of the specimen. For practical purposes E, the Young's modulus, can be assumed to be the same for compression and tension.

$$\text{Compressive stress} = \frac{\text{compressive load (N)}}{\text{area of cross section (mm}^2)} \text{ N/mm}^2.$$

A good example of compressive stress is found in building and structural work. All foundations, concrete, brick and steel columns are under compressive stress, and in the making or fabrication of welded columns and supports, the strength of welded joints in compression is of great importance.

Shearing stress

If a cube has its face fixed to the table on which it stands and a force is applied parallel to the table on one of the upper edges, this force per unit area is termed a shearing stress and it will deform the cube, as indicated by the dotted line (Fig. 1.19). The angle θ through which the cube is deformed is a measure of the shearing strain, while the shearing stress will be in N/mm^2.

This is a very common type of stress in welded construction. For example, if two plates are lapped over each other and welded, then a load applied to the plates as shown puts the welds under a shearing stress. If the load is known and also the shearing strength of the metal of the weld, then sufficient metal can be deposited to withstand the load.

A welded structure should be designed to ensure that there is sufficient area of weld metal in the joint to withstand safely the load required.

Fig. 1.19

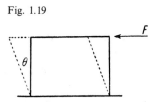

Mechanical properties of metals and the effect of heat on these properties

Plasticity may be defined as the ease with which a metal may be bent or moulded into a given shape. At ordinary temperatures, lead is one of the most plastic metals. The plasticity usually increases as temperature rises. Iron and steel are difficult to bend and shape when cold, but it becomes easy to do this when heated above red heat. Wrought iron, however, because of impurities in it, sometimes breaks when we attempt to bend it when hot (called hot shortness), and thus increase of temperatures is not always accompanied by an increase in plasticity.

Brittleness is the opposite of plasticity and denotes lack of elasticity. A brittle metal will break when a force is applied. Cast iron and high carbon steel are examples of brittle metals. The wrought iron in the above paragraph has become brittle through heating. Copper becomes brittle near its melting point, but most metals become less brittle when heat is applied. Carbon steel is an example; when cold it is extremely brittle, but can easily be bent and worked when hot. Brittle metals require care when welding them, due to the lack of elasticity.

Malleability is the property possessed by a metal of becoming permanently flattened or stretched by hammering or rolling. The more malleable a metal is, the thinner the sheets into which it can be hammered. Gold is the most malleable metal (the gold in a sovereign can be hammered into $4\,m^2$ of gold leaf, less than 0.0025 mm thick).

Copper is very malleable, except near its melting point, while zinc is only malleable between 140 and 160 °C. Metals such as iron and steel become much more malleable as the temperature rises and are readily hammered and forged.

The presence of any impurities greatly reduces the malleability, as we find that the metal cracks when it stretches.

Order of malleability when cold

(1) Gold	(3) Aluminium	(5) Tin	(7) Zinc
(2) Silver	(4) Copper	(6) Lead	(8) Iron.

Ductility is the property possessed by a substance of being drawn out into a wire and it is a property possessed in the greatest degree by certain metals. Like malleability this property enables a metal to be deformed mechanically. Metals are usually more ductile when cold, and thus wire drawing and tube drawing are often done cold, but not always.

In the wire-drawing operation, wire is drawn through a succession of tapered holes called *dies*, each operation reducing the diameter and increasing the deformation of the lattice structure. The brittleness thus increases and the wire must be softened again by a process termed annealing.

Order of ductility

(1) Gold	(3) Iron	(5) Aluminium	(7) Tin
(2) Silver	(4) Copper	(6) Zinc	(8) Lead.

Tenacity is another name for tensile strength. The addition of various substances to a metal may increase or decrease its tensile strength. Sulphur reduces the tenacity of steel while carbon increases it (see section on Tensile Strength).

Hardness is the property possessed by a metal resisting scratching or indentation. It is measured on various scales, the most common of which are: (1) Brinell, (2) Rockwell, (3) Vickers.

Table of comparative hardness

Material	Brinell	Vickers	Material	Brinell	Vickers
Lead	6	6	Brass 70/30,		
Tin	14	15	annealed	60	64
Aluminium			rolled	150	162
pure			Cast iron	150–250	160–265
annealed	19	20	Mild steel	100–120	108–130
Zinc	45	48	Stainless steel	150–165	160–180
Copper, cast	40–45	42–48			
cold worked	80–100	85–108			

Hardness decreases with rise in temperature. The addition of carbon to steel greatly increases its hardness after heat treatment, and the operations of rolling, drawing, pressing and hammering greatly affect it.

It will be noted that there is considerable latitude in the higher figures. Copper, for example, varies from 40 to 100 according to the way it is prepared. Copper is hardened by cold working, that is drawing, pressing and hammering, and this also decreases its ductility.

The tensile strength of steels can be approximately determined in N/mm^2 by multiplying their Brinell hardness figure by 3.25 for hard steels and by 3.56 for those in the soft or annealed condition.

Creep

This is the term applied to the gradual change in dimensions which occurs when a load (tensile, compressive, bending, etc.) is applied to a specimen for a long period of time. Creep generally refers to the extension which occurs in a specimen to which a steady tensile load is applied over a period of weeks and months. In these tests it is generally found that the specimen shows greater extension for a given load over a long period than for a short period and may fracture at a load much less than its usual tensile load. The effect of creep is greater at elevated temperatures and is important, as for example, in pipes carrying high-pressure steam at high superheat temperatures. In creep testing, the specimen is surrounded by a heating coil fitted with a pyrometer. The specimen is heated to a given temperature, the load is applied and readings taken of the extension that occurs over a period of weeks, a graph of the results being made. The test is repeated for various loads and at various temperatures.

Special electrodes usually containing molybdenum are supplied for welding 'creep-resisting' steels, that is, steels which have a high resistance to elongation when stresses are applied for long periods of time at either ordinary or elevated temperatures.

Fatigue

Fatigue is the tendency which a metal has to fail under a rapidly alternating load, that is a load which acts first in one direction, decreases to zero and then rises to a maximum in the opposite direction, this cycle of reversals being repeated a very great number of times. If the stress is plotted against the number of stress reversals, the curve first falls steadily and then runs almost parallel to the stress reversal axis. The stress at which the curve becomes horizontal is the fatigue limit (fig. 1.20). The load causing failure is generally much less than would cause failure if it was applied as a steady load. Many factors, such as the frequency of the applied stress, tempera-

Fig. 1.20. Number of cycles of stress reversal–stress curve.

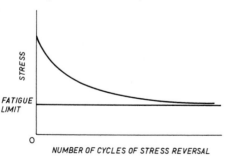

ture, internal stresses, variation in section and sharp corners leading to stress concentration, affect the fatigue limit. Methods of fatigue testing are given in Chapter 5.

Chemistry applied to welding

Elements, compounds and mixtures

All substances can be divided into two classes: (1) elements, (2) compounds.

An element is a simple substance which cannot be split up into anything simpler. For example, aluminium (Al), copper (Cu), iron (Fe), tin (Sn), zinc (Zn), sulphur (S), silicon (Si), hydrogen (H), oxygen (O) are all *elements*.

A table of the elements is given in the appendix, together with their chemical symbols.

A compound is formed by the chemical combination of two or more elements, and the property of the compound differs in all respects from the elements of which it is composed.

We have already mentioned the occurrence of matter in the form of molecules, and now it will be well to consider how these molecules are arranged among themselves and how they are made up.

If a mixture of iron filings and sand is made, we can see the grains of sand among the filings with the naked eye. This mixture can easily be separated by means of a magnet, which will attract the iron filings and leave the sand. Similarly, a mixture of sand can be separated by using the fact that salt will dissolve in water, leaving the sand. In the case of mixtures, we can always separate the components by such simple means as this (called mechanical means).

Similarly, a mixture of iron filings and powdered sulphur can be separated, either by using a magnet or by dissolving the sulphur in a liquid such as carbon disulphide, in which it dissolves readily.

Now suppose we heat this mixture. We find that it first becomes black and then, even after removing the flame, it glows like a coal fire and much heat is given off. After cooling, we find that the magnet will no longer attract the black substance which is left, neither will the liquid carbon disulphide dissolve it. The black substance is, therefore, totally different in character from the iron filings or the sulphur. It can be shown by chemical means that the iron and sulphur are still there, contained in the black substance. This substance is termed a chemical compound and is called iron sulphide. It has properties quite different from those of iron and sulphur.

Previously it has been stated that molecules can be sub-divided.

Molecules are themselves composed of atoms, and the number of atoms contained in each molecule depends upon the substance.

For example, a molecule of the black iron sulphide has been formed by the combination of one atom of iron and one atom of sulphur joined together in a chemical bond. This may be written:

$$\text{(Fe)} + \text{(S)} = \text{(Fe)}\!-\!\text{(S)}$$

The molecules of some elements contain more than one atom. A molecule of hydrogen contains two atoms, so this is written: $H_2 = O\,\underset{H\qquad H}{\overline{\qquad\qquad}}\,O$. Similarly, a molecule of oxygen contains two atoms, thus: $O_2 = O\,\underset{O\qquad O}{\overline{\qquad\qquad}}\,O$.

A molecule of copper contains only one atom, thus: $Cu = \dfrac{O}{Cu}$.

The atmosphere

Let us now study the composition of the atmosphere, since it is of primary importance in welding.

Suppose we float a lighted candle, fastened on a cork, in a bowl of water and then invert a glass jar over the candle, as shown in Fig. 1.21. We find that the water will gradually rise in the jar, until eventually the candle goes out. By measurement, we find that the water has risen up the jar $\frac{1}{5}$ of the way, that is, $\frac{1}{5}$ of the air has been used up by the burning of the candle, while the remaining $\frac{4}{5}$ of the air still in the jar will not enable the candle to continue burning. The gas remaining in the jar is nitrogen (Fig. 1.22). It has

Fig. 1.21

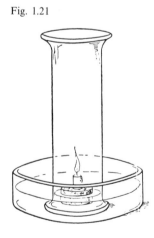

no smell, no taste, will not burn and does not support burning. The gas which has been used up by the burning candle is oxygen.

Evidently, then, air consists of four parts by volume of nitrogen to one part of oxygen. That oxygen is necessary for burning is very evident. Sand thrown on to a fire excludes the air, and thus the oxygen, and the fire is extinguished. If a person's clothes catch fire, rolling him in a blanket or mat will exclude the oxygen and put out the flames. In addition to oxygen and nitrogen the atmosphere contains a small percentage of carbon dioxide and also small percentages of the inert gases first discovered and isolated by Rayleigh and Ramsay. These gases are argon, neon, krypton and xenon. An inert gas is colourless, odourless, and tasteless, it is not combustible neither does it support combustion and it does not enter into chemical combination with other elements. Argon, which is present in greater proportion than the other inert gases, is used as the gaseous shield in 'inert gas welding' because it forms a protective shield around the arc and prevents the molten metal from combining with the oxygen and nitrogen of the atmosphere. Helium, which is the lightest of the inert gases, occurs only about 1 part in 2000 in the atmosphere but occurs in association with other natural gases in large quantities especially in the United States, where it is often used instead of argon. The various gases of the atmosphere are extracted by fractionation of liquid air.

The percentage composition, by volume, of dry air of the Earth's atmosphere is: Nitrogen 78.1, Oxygen 20.9, Argon 0.93, Neon 0.0018, Krypton 0.000 14, Xenon 0.000 008 6, Helium 0.000 52, Carbon dioxide 0.03, Hydrogen 0.000 05, Ozone 0.000 04, Methane 0.000 15, Nitrous oxide 0.000 05.

Fig. 1.22

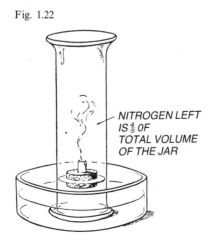

NITROGEN LEFT IS $\frac{4}{5}$ OF TOTAL VOLUME OF THE JAR

Nitrogen

Nitrogen is a colourless, odourless, tasteless gas, boiling point − 195.8 °C, which does not burn or support combustion. It is diatomic with an atomic weight of 14, and dissociates in the heat of the arc to form iron nitride, which reduces the ductility of a steel weld. For this reason it is not used as a shielding gas to any extent. It also forms nitrogen dioxide NO_2 and nitric oxide NO, which are toxic. It is widely dispersed in compound form in nitrates, ammonia and ammonium salts. It is produced by the liquefaction of air, and the considerable volumes produced by plants such as those supplying tonnage oxygen to steel plants can be used as the top pressure gas in blast furnaces and for the displacement of air in tanks, pipelines, etc.

Nitrogen is supplied in compressed form in steel cylinders of 6.2 and 4.6 m³ (220 and 165 ft³) capacity at a pressure of 13.5 N/mm², and in liquid form by bulk tankers to an evaporator which in turn feeds gas into a pipeline. (See liquid oxygen.)

Argon

This monatomic gas, chemical symbol Ar and atomic weight 18, is present in the atmosphere to the extent of about 1% and is obtained by fractional distillation from liquid air. It has no taste, no smell, is non-toxic, colourless and neither burns nor supports combustion. It does not form chemical compounds and has special electrical properties. It is extensively used in welding, either on its own or mixed with carbon dioxide or hydrogen, in the welding of aluminium, magnesium, titanium, copper, stainless steel and nickel by the TIG and MIG processes and in plasma welding of stainless steel, nickel and titanium, etc. Argon is used for the inert gas filling of electric lamps and valves, with nitrogen, and in metal refining and heat treatment, for inert atmospheres. It is supplied in compressed form in steel cylinders of 8.5 m³ or 68 m³ capacity at a pressure of 17.2 N/mm² and in liquid form delivered by road tankers which pump it directly into vacuum-insulated storage vessels as for liquid oxygen (q.v.).

Helium

Helium is an inert gas only present in the atmosphere to an extent of 0.000 052%. It is obtained from underground sources in the USA and is very much more expensive in this country than argon. It is monatomic with an atomic weight 4, is lighter than argon and is the lightest of the rare gases. Like the other inert gases it is colourless, odourless and tasteless, does not burn or support combustion, is non-toxic and does not form chemical compounds. Because of its lightness, a flow rate of 2 to $2\frac{1}{2}$ times that of argon is required to provide an efficient gas shield in inert gas welding processes. Mixed with argon it gives a range of proprietary gases for TIG and MIG welding processes.

Carbon dioxide CO_2

Carbon dioxide is now extensively used as a shielding gas in the gas shielded metal arc welding process. It is a non-flammable gas of molecular weight of 44.01, with a slightly pungent smell and is about $1\frac{1}{2}$ times as heavy as air (specific gravity relative to air is 1.53). It is soluble in water, giving carbonic acid H_2CO_3, and it can be readily liquefied, the liquid being colourless; the critical temperature (that is the temperature above which it is impossible to liquefy a gas by increasing the pressure) is $31.02\,°C$. Because its heat of formation is high it is a stable compound, enabling it to be used as a protective shield around the arc to protect the molten metal from contamination by the atmosphere, and it can be mixed with argon for the same purpose. During the CO_2 shielded metal arc process some of the molecules will be broken down or dissociated to form small quantities of carbon monoxide and oxygen. The carbon monoxide recombines with oxygen from the atmosphere to form CO_2 again and only very small quantities (the generally accepted threshold is 50 p.p.m.) escape into the atmosphere and the oxygen is removed by powerful deoxidizers in the welding wire. The gas is very much cheaper than argon; it is not an inert gas.

Carbon dioxide is formed when limestone is heated strongly in the lime kiln and also by the action of hydrochloric acid on limestone. It may be obtained as a by-product in the production of nitrogen and hydrogen in the synthesis of ammonia and also as a by-product in the fermentation process when yeast acts on sugar or starch to produce alcohol and carbon dioxide.

Large supplies for industrial use may be obtained by burning oil, coke or coal in a boiler. The steam generated can be used for driving prime movers for electricity generation and the flue gases, consisting of CO_2, nitrogen and other impurities, are passed into a washer where the impurities are removed and then into an absorber where the CO_2 is absorbed and the nitrogen thus separated. The absorber containing the CO_2 passes into a stripping column where the CO_2 is removed and water vapour set free, and is removed by a condenser. The CO_2 is then stored in a gas-holder from which it passes through a further purifying process, is then compressed in a compressor, passed through a drier and a condenser and stored in the liquid state at a pressure of 2 N/mm² at a temperature of $-18\,°C$, the storage tank being well insulated. The liquid CO_2 is then pumped into the cylinders used for welding purposes or into bulk supply tanks, or it may be further converted into the solid state (Cardice) which has a surface temperature of $-78.4\,°C$.

The use of CO_2 as a shielding gas is fully discussed in the chapter on this process and, in addition to this, the following are the main uses of the gas at the present time: in nuclear power stations, where it can be used for transference of heat from the reactor to the electricity generating unit; for the CO_2–silicate process in the foundry for core and mould making; for the soft drink trade where the gas is dissolved under pressure in the water of the mineral water or beer and gives a sparkle to the drink when the pressure is released; and in the solid state for refrigerated transport, the perishable foodstuffs being packed in heavily insulated containers with the solid CO_2, which evaporates to the gaseous state and leaves no residue.

Oxygen

In view of the importance of oxygen to the welder, it will be useful to prepare some oxygen and investigate some of its properties.

Place a small quantity of potassium chlorate in a hard glass tube (test tube) and heat by means of a gas flame. The substance melts, accompanied by crackling noises. Now place a glowing splinter in the mouth of the tube. The splinter bursts into flame and burns violently (Fig. 1.23). Oxygen is being given off by the potassium chlorate and causes this violent burning. The glowing splinter test should *not* be used for testing for an escape of oxygen from welding plant.

Oxygen is prepared on a commercial scale by one of two methods: (1) liquefaction of air; (2) electrolysis of water. In the first method air is liquefied by reducing its temperature to about $-140\,^\circ C$ and then compressing it to a pressure of 40 bar (4 N/mm²). The pressure is then reduced and the nitrogen boils off first, leaving the liquid oxygen behind. This is then allowed to boil off into its gaseous form and is compressed into steel cylinders at 175 bar (17.5 N/mm²).

Fig. 1.23. Test for oxygen, glowing splinter bursts into flame.

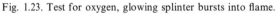

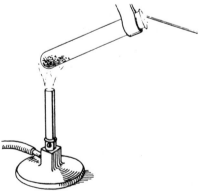

The second method is generally used when there is a plentiful supply of cheap water power for generating electricity. An electric current is passed through large vats containing water, the current entering at the anode (positive) and leaving at the cathode (negative). The passage of the current splits up the water into hydrogen and oxygen. The hydrogen is collected from the cathode and the oxygen from the anode, there being twice the volume of hydrogen evolved as oxygen. (This operation is known as electrolysis.) The gases are then dried, compressed and stored in steel containers, the hydrogen being compressed to 172 bar (17.2 N/mm^2), similar to the oxygen.

Properties of oxygen. Oxygen is a colourless gas of atomic weight 16, boiling point $-$ 183 °C, with neither taste nor smell. It is slightly soluble in water, and this slight solubility enables fish to breathe the oxygen which has dissolved.

Oxygen itself does not burn, but it very readily supports combustion, as shown by the glowing splinter which is a test for oxygen.

If a piece of red-hot iron is placed in oxygen it burns brilliantly, giving off sparks. This is caused by the iron combining with the oxygen to form an *oxide*, in this case iron oxide (Fe_3O_4).

Oxidation. Most substances combine very readily with oxygen to form oxides, and this process is termed *oxidation*.

Magnesium burns brilliantly, forming a white solid powder, magnesium oxide, i.e.:

$$\text{magnesium} + \text{oxygen} \rightarrow \text{magnesium oxide}$$
$$2Mg + O_2 \rightarrow 2MgO.$$

When copper is heated to redness in contact with oxygen copper oxide is formed:

$$\text{copper} + \text{oxygen} \rightarrow \text{copper oxide}$$
$$2Cu + O_2 \rightarrow 2CuO.$$

Similarly, phosphorus burns with a brilliant flame and forms phosphorus oxide (P_2O_5). Sulphur burns with a blue flame and forms the gas, sulphur dioxide (SO_2).

Silicon, if heated, will combine with oxygen to form silica (SiO_2), which is sand:

$$\text{silicon} + \text{oxygen} \rightarrow \text{silica or oxide of silicon}$$
$$Si + O_2 \rightarrow SiO_2.$$

Burnt dolomite, used as a refractory lining in the basic steel making process, is formed of magnesium and calcium oxides $MgO.CaO$.

When a chemical action takes place and heat is given out it is termed an exothermic reaction. The combination of iron and sulphur (p. 33), silicon and oxygen, and aluminium and iron oxide (p. 43) are examples. If heat is taken in during a reaction it is said to be endothermic. An example is the reaction which occurs when steam is passed over very hot coke. The oxygen combines with the carbon to form carbon monoxide and hydrogen is liberated, the mixture of the two gases being termed water gas, or:

$$\text{steam} + \text{carbon} \rightarrow \text{carbon monoxide} + \text{hydrogen}$$
$$H_2O + C \rightarrow CO + H_2.$$

The rusting of iron. Moisten the inside of a glass jar so that small iron filings will adhere to the interior surface and invert the jar over a bowl of water, thus entrapping some air inside the jar (Fig. 1.24).

If the surface of the water inside the jar is observed, it is seen that as time passes and the iron filings become rusty, the surface of the water rises and eventually remains stationary at a point roughly $\frac{1}{5}$ of the way up the jar. From the similar experiment performed with the burning candle it can be seen that the oxygen has been used up as the iron rusts and nitrogen remains in the jar. The rusting iron is, therefore, a process of surface oxidation.

This can further be demonstrated as follows: boil some water for some time in a glass tube (or test tube), in order to expel any dissolved oxygen, and then place a brightly polished nail in the water. Seal the open end of the tube by pouring melted vaseline down onto the surface of the water. The nail will now keep bright indefinitely, since it is completely out of contact with oxygen.

Fig. 1.24

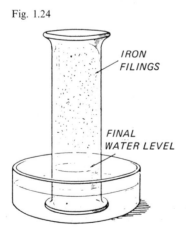

IRON FILINGS

FINAL WATER LEVEL

Oxidation, from the welder's point of view, is the union of a metal with oxygen to form an oxide, i.e.:

metal + oxygen → metallic oxide.

Oxygen reacts with metals in various ways, depending on:

(1) *The character of the metal.* Magnesium burns very completely to form magnesium oxide, while copper, aluminium and chromium form a protective oxide film on their surface at room temperature.

(2) *Temperature.* Zinc at normal temperature only oxidizes slowly on the surface, but if heated to high temperature it burns with a bright bluish-white flame, forming a white powder, zinc oxide. Nearly all base metals can be converted to their oxide by heating them in oxygen.

(3) *The amount of surface exposed.* The larger the surface area the greater the amount of oxidation.

(4) *The amount of oxygen present.* Oxidation is much more rapid, for example, in a stream of pure oxygen than in air.

(5) *Presence of other substances.* Iron will not rust if no water is present.

Let us now examine the extent to which the more important metals in welding react to oxygen.

Iron and steel. If iron is excessively heated, oxygen is absorbed and oxidation or burning takes place, forming magnetic oxide of iron:

iron + oxygen → magnetic oxide of iron

$$3Fe + 2O_2 \rightarrow Fe_3O_4.$$

There are two other oxides of iron, ferric oxide (F_2O_3) or haematite, which is one of the sources of iron from the earth, and ferrous oxide (FeO), which is a black powder which takes fire when heated in air and forms ferric oxide.

Copper is extremely resistant to atmospheric corrosion, since it forms a film of oxide on its surface. This film is very unlike rust on iron, because it protects the metal and offers high resistance to any further attack. In time the oxide becomes changed to compounds having a familiar green colour such as sulphate of copper. When copper is brightly polished and exposed to a clean, dry atmosphere it tarnishes and becomes coated with a thin film of cuprous oxide (Cu_2O). If the temperature of the copper is now raised, the amount of oxidation increases proportionally and at high temperatures the copper begins to scale. The black scale formed is cupric oxide (CuO), while underneath this is another film of cuprous oxide (Cu_2O), which has a characteristic red colour.

Aluminium has a great affinity for oxygen and is similar to copper in that it forms a protective coating (of aluminium oxide, Al_2O_3) on its surface, which protects it against further attack. The depth of the film of oxide formed will depend upon the amount of corrision, since the film adjusts itself to the amount of corrosive influences.

As the temperature increases little alteration takes place until near the melting point, when the rate of oxidation increases rapidly. It is the formation of this oxide which makes the welding of aluminium almost impossible unless a chemical (termed a flux) is used to dissolve it or an inert gas shield is used to prevent oxidation.

During the welding process, therefore, combination of the metal with oxygen may:
  (1) Produce a gaseous oxide of a metal present in the weld and thus produce blow or gas holes.
  (2) Produce oxides which, having a melting point higher than that of the surrounding metal, will form solid particles of *slag* in the weld metal.
  (3) Produce oxides which will dissolve in the molten metal and make the metal brittle and weak. (The oxide in this case may form along the boundaries of the crystals of the metal.)

Some oxides are heavier than the parent metal and will tend to sink in the molten weld. Others are lighter and will float to the top. These are less troublesome, since they are easier to remove.

Oxides of wrought iron and steel, for example, melt very much below the temperature of the parent metal and, being light, float to the surface as a scale. Thus, if care is taken in the welding process, the oxide is not troublesome.

In the case of cast iron, however, the oxide melts at a temperature above that of the metal; consequently, it would form solid particles in the weld if not removed. For this reason a 'flux' is used which combines with the oxide and floats it to the surface. In welding copper, aluminium, nickel and brass, for example, a flux must be used to remove the oxides formed (see pp. 55–7).

The two most common causes of oxidation in welding are absorption of oxygen from the atmosphere, and use of an incorrect flame with excess oxygen in gas welding.

Reduction or deoxidation. Reduction takes place when oxygen is removed from a substance. Evidently it is always accompanied by oxidation, since the substance that removes the oxygen will become oxidized.

The great affinity of aluminium for oxygen is made use of in the thermit process of welding and provides an excellent example of chemical *reduction*.

Suppose we mix some finely divided aluminium and finely divided iron oxide in a crucible or fireclay dish. Upon setting fire to this mixture it burns and great heat is evolved with a temperature as high as 3000 °C. This is due to the fact that the aluminium has a greater affinity for oxygen than the iron has, when they are hot, and as a result the aluminium combines with the oxygen taken from the iron oxide. Thus the pure iron is set free in the molten condition. The action is illustrated as follows:

$$\text{iron oxide} + \text{aluminium} \rightarrow \text{aluminium oxide} + \text{iron}$$
$$Fe_2O_3 \quad + \quad 2Al \quad \rightarrow \quad Al_2O_3 \quad + 2Fe.$$

This is the chemical action which occurs in an incendiary bomb. The detonator ignites the ignition powder which sets fire to the thermit mixture. This is contained in a magnesium–aluminium alloy case (called Elektron) which also burns due to the intense heat set up by the thermit reaction.

Since oxygen has been taken from the iron, the iron has been *reduced* or deoxidized and the aluminium is called the reducing agent. To prevent oxidation taking place in a weld, silicon and manganese are used as deoxidizers. More powerful deoxidizers such as aluminium, titanium and zirconium (triple deoxidized) are added when oxidizing conditions are more severe, as for example in CO_2 welding and also in the flux cored continuous wire feed process without external gas shield, and the deoxidizers control the quality of the weld metal.

Note. Hydrogen is an electro-positive element, while oxygen is an electro-negative element. Therefore, oxidation is often spoken of as an increase in the ratio of the electro-negative portion of a substance, while reduction is an increase in the ratio of the electro-positive portion of a substance.

Examples of:

Oxidizing agents
(1) Oxygen
(2) Ozone
(3) Nitric acid
(4) Chlorine
(5) Potassium chlorate
(6) Potassium nitrate
(7) Manganese dioxide
(8) Hydrogen peroxide
(9) Potassium permanganate

Reducing agents
(1) Hydrogen
(2) Carbon
(3) Carbon monoxide
(4) Sulphur dioxide (at low temperatures)
(5) Sulphuretted hydrogen
(6) Zinc dust
(7) Aluminium.

Note. Dry SO_2 at welding temperatures behaves as an oxidizing agent and oxidizes carbon to carbon dioxide and many metallic sulphides to sulphates.

Acetylene

Acetylene is prepared by the action of water on calcium carbide (CaC_2). The carbide is made by mixing lime (calcium oxide) and carbon in an electric arc furnace. In the intense heat the calcium of the lime combines with the carbon, forming calcium carbide and, owing to the high temperatures at which the combination takes place, the carbide is very hard and brittle. It contains about 63% calcium and 37% carbon by weight and readily absorbs moisture from the air (i.e. it is hygroscopic); hence it is essential to keep it in airtight containers. The reaction is:

calcium oxide
or quicklime + carbon → calcium carbide + carbon monoxide
$$CaO \quad + \quad 3C \quad \rightarrow \quad CaC_2 \quad + \quad CO.$$

The carbon monoxide burns in the furnace, forming carbon dioxide.

When water acts on calcium carbide, the gas acetylene is produced and slaked lime remains:

calcium carbide + water → acetylene + slaked lime
$$CaC_2 \quad + 2H_2O \rightarrow \quad C_2H_2 \quad + \quad Ca(OH)_2.$$

Acetylene is a colourless gas, slightly lighter than air, only very soluble in water, with a pungent smell largely due to impurities. It burns in air with a sooty flame but when burnt in oxygen the flame has a bright blue inner cone. It can be ignited by a spark or even by hot metal and forms explosive compounds with copper and silver so that copper pipes and fittings should never be used with it.

If compressed it is explosive but it is very soluble in acetone, which can dissolve 300 times its own volume at a pressure of 1.2 N/mm^2 or 12 bar and this is the method used to store it in the dissolved acetylene cylinders used in oxy-acetylene welding.

Liquid petroleum gas (LPG). Propane C_3H_8, butane or C_4H_{10}. Propane is a flammable gas used as a fuel gas with either air or oxygen for heating and cutting operations. Its specific gravity compared with air is 1.4–1.6 (butane 1.9–2.1) so that any escaping gas collects at ground level and an artificial stenchant is added to the gas to warn personnel of its presence since it acts as an asphyxiant. Its boiling point at a pressure of 1 atmosphere is $- 42 \,°C$ (butane $- 7 \,°C$) and the air-propane and oxy-propane flames have a greater calorific value than air-natural gas or oxy-natural gas for the same conditions of operating pressure. Flame temperature and hence cutting speeds are lower for oxy-propane than oxy-acetylene but propane is considerably cheaper than acetylene.

Note. The oxy-propane flame cannot be used for welding.

Propane burns in air to form carbon dioxide and water and is supplied in steel cylinders painted red in weights 4.8–47 kg. being sold by weight. It is also supplied by tanker to bulk storage tanks in a similar way to oxygen and nitrogen. Liquid natural gas (LNG) is similarly supplied.

Carbon

Carbon is of great importance in welding, since it is present in almost every welding operation. It is a non-metallic element, and is remarkable in that it forms about half a million compounds, the study of which is termed *organic chemistry*.

Carbon can exist in three forms. Two of these forms are crystalline, namely diamond and graphite, but the crystals of a diamond are of a different shape from those of graphite. (Carbon is found in grey cast iron as graphite.) Ordinary carbon is a third form, which is non-crystalline or *amorphous*. Carbon forms with iron the compound ferric carbide, Fe_3C, known as cementite. The addition of carbon to pure iron in the molten state is extremely important, since the character of the iron is greatly changed. Diamond and graphite are allotropes of carbon. Allotropy is the existence of an element in two or more forms.

Carbon is found in organic compounds such as acetylene (C_2H_2), petrol (C_6H_{14}), sugar $(C_{12}H_{22}O_{11})$, ethyl alcohol (C_2H_5OH), propane (C_3H_8), butane (C_4H_{10}), methane (CH_4), natural gas, etc.

Graphite used to be considered as a lead compound, but it is now known that it is a crytalline form of carbon. It is greasy to touch and is used as a lubricant and for making pencils.

The oxides of carbon. Carbon dioxide (CO_2) is heavier than air and is easily identified. It is formed when carbon is burnt in air, hence is present when any carbon is oxidized in the welding operation:

carbon + oxygen → carbon dioxide
$$C + O_2 \rightarrow CO_2.$$

It will not burn, neither will it support combustion. It turns lime water milky, and this is the usual test for it. When it dissolves in the moisture in the air or rain it forms carbonic acid, which hastens corrosion on steel (see p. 54).

Carbon monoxide is formed when, for example, carbon dioxide is passed through a tube containing red-hot carbon:

carbon + carbon dioxide → carbon monoxide
$$C + CO_2 \rightarrow 2CO.$$

Hence it may be formed from carbon dioxide during the welding process.

It is a colourless gas which burns with a blue, non-luminous flame. It is not soluble in water, has no smell and is very poisonous, producing a form of asphyxiation. Exhaust fumes from petrol engines contain a large proportion of carbon monoxide and it is the presence of this that makes them poisonous.

Carbon monoxide readily takes up oxygen to form carbon dioxide. It is thus a reducing agent and it can be made to reduce oxides of metals to the metals themselves.

The following poisonous gases may be formed during welding operations depending upon the process used, the material being welded, its coating and the electrode type:

Gas	*Example of formation*
Carbon monoxide	In CO_2 welding due to dissociation of some of the CO_2 in the heat of the arc.
Ozone, O_3	Due to oxygen in the atmosphere being converted to ozone by the ultraviolet radiation from the arc.
Phosgene, $COCl_2$	When trichloroethylene($CHCl.CCl_2$), used for degreasing is heated or exposed to ultraviolet radiation from the arc.

Non-poisonous gases such as carbon dioxide and argon act as asphyxiants when the oxygen content of the atmosphere falls below about 18%. Fumes and pollutant gases also occur during welding due, for example, to the break-up of the electrode coating in metal arc welding and the vaporization of some of the metal used in the welding process. The concentrations of these are governed by a Threshold Limit Value (TLV) and the limits are expressed in parts per million (p.p.m.).

Combustion or burning

The study of combustion is very closely associated with the properties of oxygen. When burning takes place, a chemical action occurs. If a flame is formed, the reaction is so vigorous that the gases become luminous. Hydrogen burns in air with a blue, non-luminous flame to form water. In the oxy-hydrogen flame, hydrogen is burnt in a stream of oxygen. This causes intense heat to be developed, with a flame temperature of about 2800 °C.

The oxy-coal-gas flame is very similar, as the coal gas consists of hydrogen, together with other impurities (methane, carbon monoxide and other hydrocarbons). Because of these impurities, the temperature of this flame is much lower than when pure hydrogen is used. The oxy-acetylene flame consists of the burning of acetylene in a stream of oxygen. Acetylene is composed of carbon and hydrogen (C_2H_2), and it is a gas which burns in air with a very smoky flame, the smoke being due, as in the case of a candle, to incomplete combustion of the carbon:

$$\text{acetylene} + \text{oxygen} \rightarrow \text{carbon} + \text{water}$$
$$2C_2H_2 + O_2 \rightarrow 4C + 2H_2O.$$

By using, however, a special kind of burner, we have almost complete combustion and the acetylene burns with a very brilliant flame, due to the incandescent carbon.

The oxy-acetylene welding flame

When oxygen is mixed with the acetylene in approximately equal proportions a blue, non-luminous flame is produced, the most brilliant part being the blue cone at the centre. The temperature of this flame is given, with others, in the table:

Temperatures of various flames

Oxy-acetylene	3100 °C
Oxy-butane (Calor-gas)	2820 °C
Oxy-propane (liquefied petroleum gas, LPG)	2815 °C
Oxy-methane (natural gas)	2770 °C
Oxy-hydrogen	2825 °C
Air-acetylene	2325 °C
Air-methane	1850 °C
Air-propane	1900 °C
Air-butane	1800 °C

(Metal arc: 6000 °C upwards depending on type of arc)

This process of combustion occurs in two stages: (1) in the innermost blue, luminous cone; (2) in the outer envelope. In (1) the acetylene combines with the oxygen supplied, to form carbon monoxide and hydrogen:

$$\text{acetylene} + \text{oxygen} \rightarrow \text{carbon monoxide} + \text{hydrogen}$$
$$C_2H_2 + O_2 \rightarrow 2CO + H_2.$$

In (2) the carbon monoxide burns and forms carbon dioxide, while the hydrogen which is formed from the above action combines with oxygen to form water:

$$\text{carbon monoxide} + \text{hydrogen} + \text{oxygen} \rightarrow \text{carbon dioxide} + \text{water}$$
$$CO + H_2 + O_2 \rightarrow CO_2 + H_2O.$$

The combustion is therefore complete and carbon dioxide and water (turned to steam) are the chief products of the combustion. This is shown in Fig. 1.25. If insufficient oxygen is supplied, the combustion will be incomplete and carbon will be formed.

From this it will be seen that the oxy-acetylene flame is a strong *reducing* agent, since it absorbs oxygen from the air in the outer envelope. Much of its success as a welding flame is due to this, as the tendency to form oxides is greatly decreased. For complete combustion, there is a correct amount of oxygen for a given amount of acetylene. If too little oxygen is supplied, combustion is incomplete and carbon is set free. This is known as a carbonizing or carburizing flame. If too much oxygen is supplied, there is more than is required for complete combustion, and the flame is said to be an oxidizing flame.

For usual welding purposes the neutral flame, that is neither carbonizing nor oxidizing, is required, combustion being just complete with excess of neither carbon nor oxygen. For special work an oxidizing or carbonizing flame may be required, and this is always clearly indicated.

Silicon (Si)

Silicon is an element closely allied to carbon and is found in all parts of the earth in the form of its oxide, silica (SiO_2). In its free state, silica is found as quartz and sand. Silicon is also found combined with certain other oxides of metals in the form of silicates. Silicates of various forms are often used as the flux coverings for arc-welding electrodes and are termed 'siliceous matter'.

Silicon exists either as a brown powder or as yellow-brown crystals. It combines with oxygen, when heated, to form silica, and this takes place during the conversion of iron to steel. Silicon is present, mixed in small proportions with the iron, and, when oxygen is passed through the iron in the molten state, the silicon oxidizes and gives out great heat, an exothermic reaction.

Silicon is important in welding because it is found in cast iron (0.5–3.5%) and steel and wrought iron (up to 0.1%). It is found up to 0.3% in steel casings since it makes the steel flow easily in the casting process.

Fig. 1.25. The oxy-acetylene flame.

OUTER ENVELOPE

INNER CONE

(*HERE ACETYLENE AND OXYGEN FORM CARBON MONOXIDE AND HYDROGEN IN EQUAL VOLUMES*)

(*HERE THE CARBON MONOXIDE AND HYDROGEN COMBINE WITH OXYGEN FROM THE AIR TO FORM CARBON DIOXIDE AND WATER*)

It is particularly important in the welding of cast iron, because silicon aids the formation of graphite and keeps the weld soft and machinable. If the silicon is burnt out during welding the weld becomes very hard and brittle. Because of this, filler rods for oxy-acetylene welding cast iron contain a high percentage of silicon, being known as 'silicon cast iron rods'. This puts back silicon into the weld to replace that which has been lost and thus ensures a sound weld.

By mixing silica (sand) and carbon together and heating them in an electric furnace, silicon carbide or carborundum is formed:

silica + carbon = silicon carbide + carbon monoxide
$$SiO_2 + 3C = SiC + 2CO$$

Carborundum is used for all forms of grinding operations. Silica bricks, owing to their heat-resisting properties, are used for lining furnaces.

Iron (Fe)

Iron has a specific gravity of 7.8, melts at $1530\,°C$ and has a coefficient of expansion of 0.000 0 12 per degree C.

Pure iron is a fairly soft, malleable metal which can be attracted by a magnet.

All metallic mixtures and alloys containing iron are termed ferrous, while those such as copper, brass and aluminium are termed non-ferrous.

Iron combines directly with many non-metallic elements when heated with them, and of these the following are the most important to the welder:

With sulphur it forms iron sulphide (FeS).

With oxygen is forms magnetic oxide of iron (Fe_3O_4).

With nitrogen it forms iron nitride (Fe_4N).

With carbon it forms iron carbide (Fe_3C, called cementite).

Steel, for example, is a mixture of iron and iron carbide.

Formation of metallic crystals

We have seen that atoms in solid substances take up regular geometrical patterns termed as space lattice. There are many types of space lattice but atoms of pure metals arrange themselves mainly into three of the simpler forms termed: (1) body-centred cubic, (2) face-centred cubic, (3) hexagonal close packed. These are shown in Fig. 1.26. In body-centred cubic, atoms occupy the eight corners of a cube with one atom in the centre of the cube giving a relatively open arrangement. Face-centred cubic has eight atoms at the corners of a cube with six atoms, one in the centre of each face, giving a more closely packed arrangement. Hexagonal close packed is formed by six atoms at the corners of a regular hexagon with one in the centre, placed over a similar arrangement and with three atoms in the hollows separating top and bottom layer. The student can very simply obtain the three-dimensional picture of these arrangements by using ping-pong balls to represent atoms. Copper and aluminium have a face-centred cubic lattice, magnesium a hexagonal close packed. Iron has a body-centred cubic lattice below 900°C (alpha iron, α), this changes to face-centred cubic from 900 °C to 1400 °C (gamma iron, γ), and reverts to body-centred cubic from 1400 °C to its melting point at about 1500 °C (delta iron, δ). These different crystalline forms are allotropic modifications.

When a liquid (or pure molten) metal begins to solidify or freeze, atoms begin to take up their positions in the appropriate lattice at various spots or nuclei in the molten metal, and then more and more atoms add themselves to the first simple lattice, always preserving the ordered arrangements of the lattice, and the crystals thus formed begin to grow like the branches of a tree and, from these arms, other arms grow at right angles, as shown in Fig. 1.27. Eventually these arms meet arms of neighbouring crystals and no further growth outwards can take place. The crystal then increases in size,

Fig. 1.26. Types of crystals.

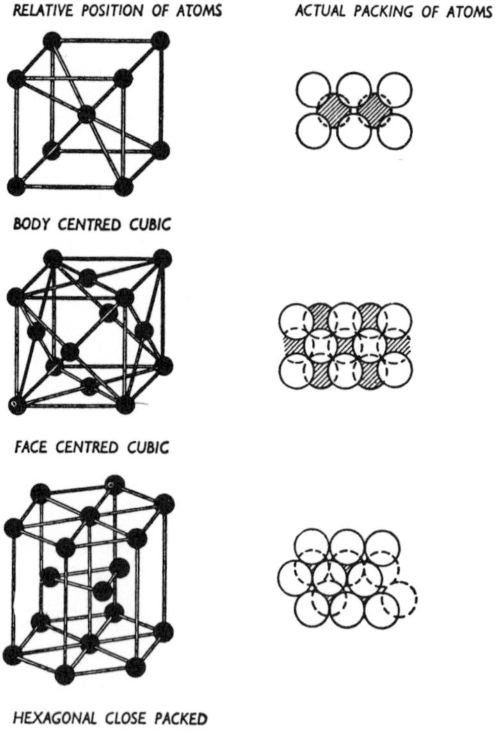

RELATIVE POSITION OF ATOMS   ACTUAL PACKING OF ATOMS

BODY CENTRED CUBIC

FACE CENTRED CUBIC

HEXAGONAL CLOSE PACKED

within its boundary, forming a solid crystal, and the junction where it meets
the surrounding crystals becomes the crystal of grain boundary. Its shape
will now be quite unlike what it would have been if it could have grown
without restriction; hence it will have no definite shape.

If we examine a pure metal structure under a microscope we can clearly
see these boundary lines separating definite areas (Fig. 1.28), and most pure
metals have this kind of appearance, it being very difficult to tell the

Fig. 1.27

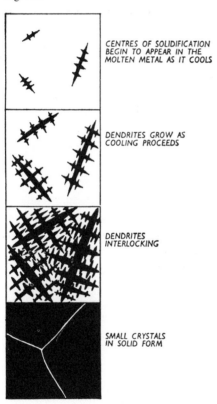

CENTRES OF SOLIDIFICATION
BEGIN TO APPEAR IN THE
MOLTEN METAL AS IT COOLS

DENDRITES GROW AS
COOLING PROCEEDS

DENDRITES
INTERLOCKING

SMALL CRYSTALS
IN SOLID FORM

Fig. 1.28

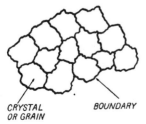

CRYSTAL   BOUNDARY
OR GRAIN

difference between various pure metals by viewing them in this way. If impurities are present, they tend to remain in the metal which is last to solidify and thus appear between the arms of the dendrites along the grain boundaries.

This method of crystallization is termed *dendritic crystallization* and the above crystal is termed a dendrite (Fig. 1.29). It can be observed when frost forms on the window pane, and this gives a good illustration of the method of crystal formation, since the way in which the arms of the dendrite interlock can clearly be seen. (The frost, however, only forms on a flat surface, while the metal crystal forms in three dimensions.) Crystals which are roughly symmetrical in shape are termed equi-axed.

Recrystallization commences at a definite temperature, and if the temperature is increased greatly above this, the grains become much larger in size, some grains absorbing others. Also, the longer that the metal is kept in the heated condition the larger the grains grow. The rate at which cooling occurs in the case of metals also determines the size of the grains, and the slower the cooling the larger the grains (see grain growth, Fig. 2.17). Large, coarse crystals or grains have a bad effect on the mechanical properties of the metal and decrease the strength. If heat conditions are suitable crystals may grow in one direction, then being long and narrow. These are termed columnar crystals. Fig. 2.22 shows a mild steel arc weld run on a steel plate. There is a thin layer of chill cast crystals on the upper surface of the weld since they cooled quickly in contact with the atmosphere and have had no time to grow. They have been left out for the sake of clarity. Below these are columnar crystals growing towards the centre of greatest heat. Lower still are equi-axed crystals showing grain growth, while below these is a region of small crystals where recrystallization has occurred.

Cold work distorts the crystals in the direction of the work (see Fig. 2.22).

The size of the grains, therefore, depends on:

(1) The type of metal.
(2) The temperature to which the metal has been raised.
(3) The length of time for which the metal is kept at high temperature.
(4) The rate of cooling.

Crystals of alloys

If one or more metals is added to another in the molten state, they mix together forming a solution, termed an *alloy*.

When this alloy solidifies:

(1) It may remain as a solution, in which case we get a crystal structure similar to that of pure metal.

Fig. 1.29
(*a*) Fir tree (dendritic) crystals in the shrinkage cavity of a large carbon steel casting. The crystals have grown mainly in one direction, but the growth of the lateral arms at right angles to the main axes is clearly seen. ($\frac{1}{2}$ actual size.)
(*b*) Portion of a nickel chrome molybdenum steel ingot showing interlocking of the dendrites. In this case crystallization has started from a series of centres, and the growth of any one dendritic crystal has been restricted by the presence of neighbouring crystals.

(a)

(b)

(2) The two metals may tend to separate out before solidifying, in which case the crystal structure will be a mixture of the crystals of the two metals, intimately mixed together.

Copper–nickel and chromium–iron are examples of the first kind of crystal formation, while lead–tin and copper–zinc are examples of the second kind.

Welding has a very great effect on the structure and crystal form of metals, and the above brief study will enable the reader to have a clearer understanding of the problem.

Metallic alloys and equilibrium diagrams are dealt with in much greater detail in Chapter 3.

Effects of corrosion on welds in steel

Corrosion is a chemical action on a metal, resulting in the conversion of the metal into a chemical compound.

The rusting of iron, which we have considered, is a good example. In the presence of air and water, the iron eventually changes into oxides and hydroxides of iron and then into hydrated carbonates.

In addition to this type of attack, the matter which is suspended in the atmosphere also assists corrosion. The very small proportion of carbon dioxide in the atmosphere becomes dissolved in the rain, forming very weak carbonic acid, and this attacks steel, again forming carbonates. In and near large towns the atmosphere contains a very much larger proportion of suspended matter than in the country. Smoke and fumes contain, among other compounds, sulphur dioxide, which again dissolves in rain to form sulphurous acid. This is oxidized into dilute sulphuric acid, which again attacks the steel, the attack being much stronger than in the case of the carbonic acid.

Near the coast, the salt in the atmosphere forms hydrochloric acid and caustic soda, and severe corrosion occurs in these areas.

In addition to this direct form of chemical attack there is a second type of attack which is at first not so apparent. When two different metals are placed in a conducting liquid, such as a dilute acid or alkali, an electric cell or battery is formed, one of the metals becoming electro-positive, while the other becomes electro-negative. The difference of electrical pressure or voltage between these two plates will depend upon the metals chosen.

In the case of a welded joint, if the weld metal is of different structure from the parent metal, we have, if a conducting liquid is present, an electric cell, the plates of which are connected together or short-circuited. The currents which flow as a result of this are extremely minute, but nevertheless they greatly accelerate corrosion. This effect is called electrolysis, and its harmful effects are now well known.

The deposited metal in the weld is never of the same composition as the parent metal, although it may have the same properties physically. In the welded region, therefore, dissimilar metals exist, and in the presence of dilute carbonic acid from the atmosphere (or dilute sulphuric acid as the case may be) electrolytic action is set up and the surfaces of the steel become pitted. Now if the weld metal is electro-positive to the parent metal, the weld metal is attacked, since it is the electro-positive plate which suffers most from the corrosive effect. On the other hand, if the parent steel plate is electro-positive to the weld metal, the plate is attacked and since its surface area is much larger than that of the weld, the effect of corrosion will be less than if the weld had been attacked.

Thus, weld metal should be of the same composition throughout its mass to prevent corrosion taking place in the weld itself. It should also be electro-negative to the parent metal to prevent electrolytic action causing pitting of its surface, and it must resist surface oxidation at least as well as the parent metal.

Fluxes

Oxy-acetylene welding

Most metals in their molten condition become oxidized by the absorption of oxygen from the atmosphere. For example, aluminium always has a layer of aluminium oxide over its surface at normal temperature, and has a very great affinity for oxygen. To make certain that the amount of oxidation is kept a minimum, that any oxides formed are dissolved or floated off, and that welding is made as easy and free from difficulties as possible, fluxes are used. Fluxes, therefore, are chemical compounds used to prevent oxidation and other unwanted chemical reactions. They help to make the welding process easier and ensure the making of a good, sound weld.

The ordinary process of soldering provides a good example. It is well known that it is almost impossible to get the solder to run on to the surface to be soldered unless it is first cleaned. Even then the solder will not adhere uniformly to the surface. If now the surface is lightly coated with zinc chloride or killed spirit (made by adding zinc to hydrochloric acid or spirits of salt until the effervescing action ceases), the solder runs very easily wherever the chloride has been. This 'flux' has removed all the oxides and grease from the surface of the metal by chemical action and presents a clean metal surface to be soldered. This makes the operation much easier and enables a much better bond with the parent metal to be obtained. Fluxes used in oxy-acetylene welding act in the same way. Flux-covered rods are now available for bronze and aluminium welding.

Brass and bronze

A good flux must be used in brass or bronze welding, and it is usual to use one of the borax type, consisting of sodium borate with other additions. (Pure borax may be used.) The flux must remove all oxide from the metal surfaces to be welded and must form a protective coating over the surfaces of the metal, when they have been heated, so as to prevent their oxidation. It must, in addition, float the oxide, and the impurities with which it has combined, to the top of the molten metal.

Aluminium and aluminium alloys

The flux must chemically remove the film of aluminium oxide (melting point over 2000 °C) and must float any impurities to the surface of the molten pool. A typical flux contains, by weight, lithium chloride 0–30%, potassium fluoride 5–15%, potassium chloride 0.6% and the remainder sodium chloride. The fluxes are very hygroscopic (absorb moisture readily) and should always be kept in an airtight container when not in use. They are very corrosive and after welding the work should be well scrubbed in hot water or treated with a 5% solution of nitric acid in water to remove all traces of flux.

The use of these fluxes has now been largely superceded by the use of the TIG and MIG processes for the welding of aluminium, using inert gases and the arc for dispersal of the oxide films.

Cast iron

When welding wrought iron and mild steel the oxide which is formed has a lower melting point than the parent metal and, being light, floats to the surface as a scale which is easily removed after welding. No flux is, therefore, required when welding mild steel or wrought iron.

In the case of cast iron, oxidation is rapid at red heat and the melting point of the oxide is *higher* than that of the parent metal, and it is, therefore, necessary to use a flux which will combine with the oxide and also protect the metal from oxidation during welding. The flux combines with the oxide and forms a slag which floats to the surface and prevents further oxidation. Suitable fluxes contain sodium, potassium or other alkaline borates, carbonates, bicarbonates and slag-forming compounds.

Copper

Copper may be welded without a flux, but many welders prefer to use one to remove surface oxide and prevent oxidation during welding. Borax is a suitable flux, and its only drawback is that the hard, glass-like scale of copper borate, which is formed on the surface after welding, is hard to remove. Special fluxes, while consisting largely of borax, contain other substances which help to prevent the formation of this hard slag.

To sum up, we may state, therefore, that fluxes are used:

(1) To reduce oxidation.

(2) To remove any oxide formed.

(3) To remove any other impurities.

Because of this, the use of a flux:

(1) Gives a stronger, more ductile weld.

(2) Makes the welding operation easier.

It is important that *too much* flux should never be used, since this has a harmful effect on the weld.

Manual metal arc flux-covered electrodes

If bare wire is used as the electrode in MMA welding many defects are apparent. The arc is difficult to strike and maintain using d.c.; with a.c. it is extremely difficult. The resulting 'weld' lacks good fusion, is porous, contains oxides and nitrides due to absorption of oxygen and nitrogen from the air, and as a result the weld is brittle and has little strength. To remedy these defects electrodes are covered with chemicals or fluxes which:

(1) Enable the arc to be struck and maintained easily on d.c. or a.c. supplies.

(2) Provide a shield of gases such as hydrogen or carbon dioxide to shield the molten metal in its transference across the arc and in the molten pool in the parent plate from reacting with the oxygen and nitrogen of the atmosphere to form oxides and nitrides, which are harmful to the mechanical properties of the weld.

(3) Provide a slag which helps to protect the metal in transit across the arc gap when the gas shield is not voluminous, and which when solidified protects the hot metal against oxidation and slows the rate of cooling of the weld; also slag-metal reactions can occur which alter the weld metal analysis.

Also alloying elements can be added to the coverings in which case the core wire analysis will not match the weld metal analysis.

We have seen that an oxide is a compound of two elements, one of which is oxygen. Many oxides are used in arc welding fluxes, examples of which are: silicon dioxide SiO_2 (A), manganous oxide MnO (B), magnesium oxide MgO (B), calcium oxide CaO (B), aluminium oxide Al_2O_3 (Am), barium oxide BaO (B), zinc oxide ZnO (Am), ferrous oxide FeO (B). Oxides may be classified thus: acidic, basic and amphoteric, indicated by A, B and Am above (other types are dioxides, peroxides, compound oxides and neutral oxides). An acidic oxide reacts with water to form an acid thus:

$$SO_2 \text{ (sulphur dioxide)} + H_2O = \text{sulphurous acid } (H_2SO_3).$$

In the case of silicon dioxide SiO_2, which is insoluble in water, it reacts similarly with fused sodium hydroxide ($NaOH$) to form sodium silicate and water thus:

$$SiO_2 + 2\ NaOH = Na_2SiO_3 + H_2O.$$

Basic oxides interact with an acid to form a salt and water only thus: calcium oxide (CaO) reacts with hydrochloric acid (HCl) to form calcium chloride ($CaCl_2$) and water:

$$CaO + 2\ HCl = CaCl_2 + H_2O.$$

Amphoteric oxides can exhibit either basic or acidic properties. Aluminium oxide (Al_2O_3) reacts with dilute hydrochloric acid as a basic oxide thus:

$$Al_2O_3 + 6\ HCl = 2AlCl_3 + 3H_2O,$$

but as an acidic oxide it reacts with sodium hydroxide thus:

$$Al_2O_3 + 2\ NaOH = 2\ NaAlO_2 + H_2O.$$

When oxides are mixed to form fluxes the ratio of the basic to acidic oxides is termed the basicity and is important, as for example in the fluxes used for submerged arc welding (q.v.) where the flux must be carefully chosen in conjunction with the electrode wire to give the desired mechanical properties to the weld metal.

To illustrate the action of the flux-covered electrode we may consider the reaction between a basic oxide such as calcium oxide (CaO) and an acidic oxide such as silicon dioxide (SiO_2). With great application of heat these will combine chemically to form calcium silicate, which is a slag, thus:

$$\text{calcium oxide} + \text{silicon dioxide} \rightarrow \text{calcium silicate}$$
$$CaO \quad + \quad SiO_2 \quad \rightarrow \quad CaSiO_3.$$

Similarly if we use iron oxide (Fe_2O_3) and silicon dioxide (SiO_2) in the covering of the electrode, in the heat of the arc they will combine chemically to form iron silicate, which floats to the top of the molten pool as a slag, protects the hot metal from further atmospheric oxidation and slows down the cooling rate of the weld.

$$\text{iron oxide} + \text{silicon dioxide} = \text{iron silicate}$$
$$Fe_2O_3 \quad + \quad 3SiO_2 \quad = Fe_2(SiO_3)_3$$

The most common slag-forming compounds are rutile (TiO_2), limestone ($CaCO_3$), ilmenite ($FeTiO_3$), iron oxide (Fe_2O_3), silica (SiO_2), manganese oxide (MnO_2) and various aluminium silicates such as felspar and kaolin, mica and magnesium silicates.

Deoxidizers such as ferrosilicon, ferromanganese and aluminium are also added to reduce the oxides that would be formed in the weld to a negligible amount.

The chemical composition of the covering also has an effect on the electrical characteristics of the arc. Ionizers such as salts of potassium are added to make striking and maintaining the arc easier, and for arcs of the same length there is a higher voltage drop when a coating releases hydrogen as the shielding gas than when one releases cabon dioxide. For a given current this higher voltage drop gives greater energy output from the arc, and hydrogen releasing coatings are usually cellulosic, giving a penetrating arc, thin slag cover and quite an amount of spatter. Other coatings are discussed in detail in the section in Volume 2, Chapter 3 under the heading 'Classification of covered electrodes for the manual metal arc welding of carbon and carbon–manganese steels'.

For the arc welding of bronze (also copper and brass) the flux must dissolve the layer of oxide on the surface and, in addition, must prevent the oxidation of the metal by providing the usual sheath. These coatings contain fluorides (cryolite and fluorspar) and borates and the rods are usually operated on the positive pole of a direct current supply.

The flux of the aluminium rod is a mixture of chlorides and fluorides, as for oxy-acetylene welding of aluminium. It acts chemically on the oxide, freeing it, and this enables it to be floated to the surface of the weld. It is corrosive and also tends to absorb moisture from the air; hence the weld should be well cleaned with hot water on completion, while the electrodes should be stored in a dry place. In fact, all electrodes should be kept very dry, since the coatings tend to absorb moisture and the efficiency of the rod is greatly impaired if the covering is damp.

The manual metal arc welding of aluminium and its alloys has been largely superseded by the inert gas shielded-metal arc processes, TIG and MIG.

Materials used for electrode coatings

(1) Rutile. Rutile is a mineral obtained from rutile-bearing sands by suction dredging. It contains about 88–94% of TiO_2 and is probably the most widely used material for electrode coatings. Ilmenite is a naturally occurring mineral composed of the oxides of iron and titanium $FeTiO_3$ (FeO, TiO_2) with about 45–55% TiO_2. After separation of impurities it is ground to the required mesh size and varies from grey to brown in colour.

(2) Calcium carbonate or limestone is the coating for the basic coverings of electrodes. The limestone is purified and ground to required mesh size. The slag is very fluid and fluorspar is added to control fluidity. The deposited metal is very low in hydrogen content.

(3) Fluorspar or fluorite is calcium fluoride and is mined, separated from impurities, crushed, screened, and ground, and the ore constituents are

separated by a flotation process. Too great an addition of this compound to control slag fluidity affects the stability of the a.c. arc.

(4) Solka floc is cellulose acetate and is prepared from wood pulp. It is the main constituent of class 1 cellulose electrodes. Hydrogen is given off when it decomposes under the heat of the arc so that there is a large voltage drop and high power giving deep penetration. Arc control is good but there is hydrogen absorption into the weld metal.

(5) Felspar is an anhydrous silicate of aluminium associated with potash, soda, or calcium, the potash felspar being used for electrode coatings. It is used as a flux and a slag-producing substance. The potash content stabilizes the a.c. arc and it is generally used in association with the rutile and iron oxide–silica coatings. The crystalline ore is quarried and graded according to impurities, ground, and the powder finally air-separated. The binders in general use for the materials composing electrode coatings are silicates of potassium and sodium.

(6) Ball clay. A paste for an electrode coating must flow easily when being extruded and must hold liquid present so that it will not separate out under pressure; also the freshly extruded electrode must be able to resist damage when in the wet or green conditon. Because of the way in which its molecules are arranged ball clay gives these properties to a paste and is widely used in those classes of electrodes in which the presence of hydrogen is not excluded. Found in Devon and Cornwall, it is mined, weathered, shredded, pulverized and finally sieved.

(7) Iron powder is added to an electrode coating to increase the rate of metal deposition. In general, to produce the same amount of slag with this powder added to a coating it generally has to be made somewhat thicker. To produce the iron powder, pure magnetic oxide of iron is reduced to cakes of iron in a bed of carbon, coke and limestone. The 'sponge cakes' which are formed are unlike ordinary iron in that they can be pulverized to a fine powder which is then annealed. Electrodes may contain up to 50% of iron powder.

(8) Ferromanganese is employed as a deoxidizer as in steel-making to remove any oxide that has formed in spite of the arc shield. It reacts with iron oxide to form iron and manganese dioxide, which mixes with the slag.

(9) Mica is a mineral found widely dispersed over the world. It is mined and split into sheets. The larger sizes are used for electrical purposes such as commutator insulation and the smaller pieces are ground into powder form. It is used in electrode coatings as a flux and it also assists the extrusion and gives improved touch welding properties with increase in slag volume.

(10) Sodium alginate is extracted from certain types of seaweed. It is used

in electrode coatings because when made into a viscous paste, it assists extrusion and is especially useful when the coating contains a large proportion of granules.

The flux coating on electrodes may be applied in one of the following ways:*
(1) Solid extrusion.
(2) Extrusion with reinforcement.
(3) Dipping.

(1) Solid extrusion. This is the way in which most of the present-day electrodes are produced. The flux, in the form of a paste, is forced under pressure around the wire core. The thickness of the flux covering can be accurately controlled and is of even thickness all round the wire core, the method being suitable for high speed production. This covering, however, will not stand up to very rough handling, nor to bending (as is sometimes required when welding awkwardly placed joints) since the covering flakes off.

(2) Extrusion with reinforcement. In this method the reinforcement enables the covering to withstand more severe handling and bending without flaking. The reinforcement may be:
   (*a*) An open spiral of yarn wound on the rod, the space between the spiral coils being filled with extruded flux.
   (*b*) A close spiral of yarn wound over a solid extrusion of flux applied first to the wire coil, and this flux covering strengthened by a yarn or by a single or double helical wire winding.

(3) Dipping. The dipping process has been largely superseded by the other two methods. Certain rods having special applications are, however, still made by this process. Repeated dippings are used to give thicker coatings.

* Methods 2 and 3 are included for reference purposes only. Method 1 is used almost exclusively at the present time. See Vol. 2. Appendix 7.

2

Metallurgy

Production and properties of iron and steel

Before proceeding to a study of iron and steel it will be well to understand how they are produced.

Iron is found in the natural form as iron ores. These ores are of four main types:

(1) Haematite, red or brown Fe_2O_3 containing 40–60% iron.

(2) Magnetite or magnetic oxide of iron, Fe_3O_4, containing up to 70% iron.

(3) Limonite, a hydrated ore, $Fe_2O_3 \cdot 3H_2O$, containing 20–50% iron.

(4) Siderite, a carbonate, $FeCO_3$, with iron content 20–30%.

Limonite and siderite are termed lean ores since they are so low in iron. The ore found in England in Lincolnshire, Northamptonshire, Leicestershire and Oxfordshire is one of low iron content and is generally obtained by opencast working.

Iron ore, as mined, contains appreciable amounts of earthy waste material known as gangue, and if this were fed into the furnace with the ore, more fuel would be consumed to heat it up and it would reduce the furnace capacity. Ores are washed, or magnetically separated in the case of the magnetic ores, to remove much of this waste material. They are roasted or calcined to drive off the moisture and carbon dioxide and to remove some of the sulphur by oxidation to sulphur dioxide, and crushed to bring the lumps to a more uniform size.

Agglomeration of ores

Finer particles of ores (fines) cannot be fed into the furnace because they would either be blown out or would seal up the spaces in the burden (coke, ore and flux) necessary for the passage of the blast. The smaller particles can be made to stick together or agglomerated either by sintering or pelletizing.

Sintering. The materials are chiefly iron ore fines, blast furnace flue dust, limestone and/or dolomite and coke breeze or fine anthracite as fuel. They are mixed, moistened and loaded on to a moving grate consisting of pallets through which air can circulate. The mixture is ignited by gas or oil jets and burns, sucking air through the bed. The sinter is tipped from the end of the moving gate, large lumps being broken up by a breaker.

Pelletizing. The ore is usually wet concentrates made into a thick slurry to which a small amount of bentonite is added. This is then balled by feeding it into a slowly rotating drum inclined at 5–10° to the horizontal. The green balls are then fed into a vertical shaft furnace or onto a travelling grate as in sintering where they are dried, fired and cooled.

The blast furnace

The furnace is a vertical steel stack lined with refractories. Charging is done at the top and pig iron and slag are tapped from the bottom (Fig. 2.1). Large volumes of gases (including carbon monoxide) are

Fig. 2.1. The blast furnace.

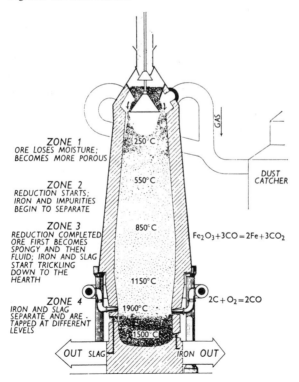

ZONE 1
ORE LOSES MOISTURE;
BECOMES MORE POROUS

ZONE 2
REDUCTION STARTS;
IRON AND IMPURITIES
BEGIN TO SEPARATE

ZONE 3
REDUCTION COMPLETED
ORE FIRST BECOMES
SPONGY AND THEN
FLUID; IRON AND SLAG
START TRICKLING
DOWN TO THE
HEARTH

ZONE 4
IRON AND SLAG
SEPARATE AND ARE
TAPPED AT DIFFERENT
LEVELS

GAS

DUST
CATCHER

250°C
550°C
850°C $Fe_2O_3 + 3CO = 2Fe + 3CO_2$
1150°C
1900°C $2C + O_2 = 2CO$
1500°C

OUT SLAG   IRON OUT

evolved during operation of the furnace and are burnt in stoves which provide the heat to raise the temperature of the air blast to about 1350 °C. This reduces the amount of coke required because combustion speed is increased and thus efficiency is increased and there is a reduction in the sulphur content of the pig iron. Four-fifths of the air in the blast is nitrogen, which takes no part in the process yet has to be raised in temperature. By enriching the blast with oxygen (up to 30%) the nitrogen volume is reduced and the efficiency increased.

During the operation of the furnace the burning coke produces carbon monoxide, which reduces the ore to metal. This trickles down to the bottom of the furnace where the temperature is highest. The limestone is decomposed into lime (calcium oxide), which combines with the silica in the gangue to form a slag, calcium silicate, and the iron begins to take in carbon. Slag is tapped from the upper notches and pig iron from the lower (Fig. 2.1).

Blast furnaces in Europe, the United States and Japan are becoming increasingly larger with greater productivity. Modern furnace can use oil fuel injection, top pressure, high blast temperatures, oxygen enrichment and pre-reduced burden in the quest for greater economy in energy and increased productivity.

Because of the large volume of these furnaces (4600 m³) it is difficult to distribute the reducing gases evenly throughout the burden, so ore size is carefully graded, strong coke is used, and equalization is done by high-top-pressure nitrogen, which reduces the velocity of the gas in the lower regions of the furnace, keeping the gas in longer contact with the burden and allowing it to ascend more uniformly thus achieving more efficient reduction.

Furnace construction can be by the stack being welded on to a ring girder which is supported on four columns or there can be a free-standing stack within a structure of four columns. The refractories are carbon and carbon with graphite for the tuyères and hearth, and aluminium oxide (alumina) for the stack. Cooling is by forced-draught air or by water for the underhearth, and flat copper coolers or staves are used for the stack with open- or closed-circuit cooling water systems.

Typical burdens are 80% sinter, 20% ore or 60/40% sinter, 40/60% pellets; coke and burden are screened, weighed and delivered to the furnace on a charging conveyor, the charging system being either double bell or bell-less with a distribution chute. Furnace charging may be done automatically and can be fully computerized. The gas cleaning plant incorporates a dust catcher and water scrubber.

Direct reduction of iron ore

As alternatives to the blast furnace method of producing iron from its ore, other processes can be used, not dependent upon the use of coke. The ore is converted into metallized pellets or sponge iron by removing the oxygen from the iron to leave metallic iron. The amount of metallic iron produced from a given quantity of ore is termed the degree of metallization and is the ratio of the amount of metallic iron produced to the total iron in the ore. The iron left after the removal of the oxygen has a honeycomb structure and is often termed sponge iron.

Direct reduction (DR) processes may be classed according to the type of fuel used: either gaseous hydrocarbons using reduced gases produced by reforming from natural gas (methane); or solid fuel such as coal or coke breeze. The gaseous fuel type can be a vertical retort (Hyl), vertical shaft furnace (Midrex) or a fluidized bed (HIB) while the solid fuel type uses rotary hearths or kilns (SL/RN, Krupp). A high degree of purity of ore is required for sponge iron because gangue is not removed at the iron-making stage but later in the steel-making process, so that the more gangue present, the less the efficiency of the process. At present pelletized concentrates are used but screened natural lump ore of similar purity can now be used as processing difficulties have been overcome.

Typical of the gaseous type of direct reduction plant installed by British Steel is the Midrex, using natural gas as the reductant. The natural gas is steam-reformed to produce carbon monoxide and hydrogen thus:

methane steam   carbon monoxide   hydrogen
$$CH_4 + H_2O + \text{heat} \rightarrow CO + 3H_2.$$

Other hydrocarbons such as naphtha or petroleum can also be used as reductants.

Considering haematite, Fe_2O_3, as the ore, the carbon monoxide and hydrogen which are both reducing gases act as follows on the ore, reducing it to metallic iron of spongy appearance, the reduction taking place above 800 °C.

$$Fe_2O_3 + 3CO \rightarrow 2Fe + 3CO_2 + \text{heat (exothermic).}$$
$$Fe_2O_3 + 3H_2 + \text{heat} \rightarrow 2Fe + 3H_2O \text{ (endothermic).}$$

Cold oxide pellets are fed by successive additions into the top of the vertical shaft furnace up which flows a counter-current of heated reducing gas (carbon monoxide and hydrogen). Metallization occurs and the metallized pellets are taken from the bottom of the furnace so that the process is continuous and economical in labour, achieving a metallization of 92–95%, the off-gases being recovered and recycled (Fig. 2.2).

DR iron is used chiefly in the electric arc and basic oxygen furnaces and it is evident that the future of this gaseous type of reduction depends upon a continuing supply of natural gas at a price competitive with that of coke.

Cast iron

Pig iron from the blast furnace is not refined enough for making castings, so in the foundry the iron for casting is prepared as follows: a coke fire is lit at the bottom of the small blast furnace or cupola and then alternate layers of pig iron (broken up into pieces) and scrap and coke together with small quantities of limestone as flux (for purifying and deslagging) are added. When the mass has burnt up, the blast is turned on

Fig. 2.2. Direct reduction of iron ore by the Midrex process.

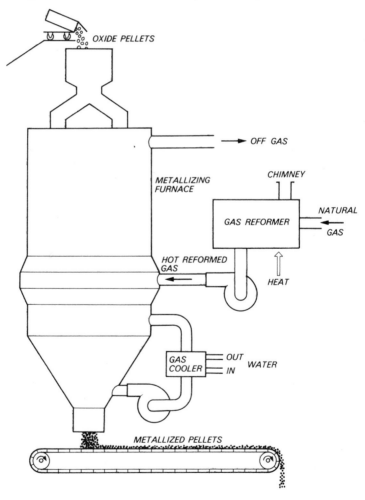

and the molten iron (melting point 1130 °C) flows to the bottom of the furnace from where it is tapped into ladles or moulds direct. Cast iron is relatively cheap to produce and its melt fluidity gives excellent casting properties. It is an alloy of many elements, and an average composition is: iron 95–98%, carbon 3–4%, silicon below 3%, sulphur below 0.2%, phosphorus below 0.75%, manganese below 1%.

The carbon exists in two forms: chemically combined carbon, and free carbon simply mixed with the iron and known as graphite. The grey look of a fracture of grey cast iron is due to this graphite, which may be from 3 to $3\frac{1}{2}$%, while the chemically combined carbon may range from 0.5 to 1.5%. As the amount of combined carbon increases, so do the hardness and brittleness increase, and if cast iron is cooled or chilled quickly from a very high temperature, the amount of combined carbon is increased, and the free carbon is reduced. As a result this type of cast iron is more brittle and harder than grey iron, and since it has a white appearance at a fractured surface, it is termed white cast iron. This has from 3 to 4% of carbon chemically combined.

Cast iron possesses very low ductility, and for this reason it presents difficulty in welding because of the strains set up by expansion and contraction tending to fracture it.

The properties of cast iron can be modified by the addition of other elements. Nickel gives a fine grain and reduces the tendency of thin sections to crack, while chromium gives a refined grain and greatly increases the resistance to wear. The addition of magnesium enables the graphite normally present in flake form to be obtained in spheroidal form. This SG (spheroidal graphite) cast iron is more ductile than ordinary cast iron (see p. 79).

Wrought iron

Wrought iron is now of historical interest only and is very difficult to obtain. It is manufactured from pig iron by the pudding process which removes impurities such as carbon, sulphur and phosphorus leaving nearly pure iron. A typical analysis is: iron 99.5–99.8%, carbon 0.01–0.03%, silicon 0–0.1%, phosphorus 0.04–0.2%, sulphur 0.02–0.04%, and manganese 0–0.25%.

When fractured, wrought iron shows a fibrous or layered structure but will bend well and is easily worked when hot. It does not harden on cooling rapidly and can be welded in the same way as mild steel.

Steel-making

Pig iron consists of iron together with 3–4% carbon, present either in the combined form as iron carbide or in the free form as graphite, the composition depending upon the type of iron ore used. In addition it contains other elements, the chief of which are manganese, silicon, sulphur and phosphorus. By oxidation of the carbon and these other elements the iron is converted into steel, the composition of which will have the required carbon and manganese percentage with a very small amount of sulphur and phosphorus (e.g. sulphur 0.02% max. and phosphorus 0.03% max.) for a typical welding-quality steel. There are two processes in steel-making, acid and basic, and they differ in the type of slag produced and in the refractory furnace lining. In the acid process, low-sulphur and -phosphorus pig iron, rich in silicon, produces an acid slag (silica) and the furnace is lined with silica refractories to prevent reaction of the lining with the slag.

In the basic process, which is largely used nowadays, phosphorus-rich pig irons can be treated with lime (a base) added to the slag to reduce the phosphorus content, and the slag is now basic. The refractory lining of the furnace must now be of dolomite (CaO, MgO) or magnesite (MgO) to prevent reaction of the slag with the furnace lining. This basic process enables widely distributed ores, with high phosphorus and low silica content, to be used.

Much steel was formerly made by the Bessemer process (acid or basic), in which an air blast is blown through the charge of molten iron contained in a steel converter lined with refractories, oxidizing the impurities, with the exothermic reaction providing the necessary heat so that no external heat source is required. The large volume of nitrogen in the air blown through the charge wastes much heat and in addition nitrogen forms nitrides in the steel reducing the deep-drawing properties of the steel, so the process is now little used.

In the *Open Hearth* (acid or basic) process, now obsolete, the heat required to melt and work the charge is obtained from the burning of a producer gas-and-air mixture over the hearth. Both gas and air are heated to a high temperature (1200 °C) by passing them through chambers of checkered brickwork in which brick and space alternate. In order that the process is continuous a regenerative system is used (Fig. 2.3). There are two sets of chambers each, for gas and air. While the gas and air are being heated in their passage through one pair of chambers, the high-temperature waste gases are pre-heating the other pair ready for the change-over. Pig iron and scrap are fed into the furnace by mechanical chargers, the pig iron being fed in the molten condition if the steel-making furnace is near the blast furnace as in integrated plants. The charging doors are on one side and the tapping hole on the other, the hearth capacity being 30–200 tonnes.

When the charge is molten, iron ore is added and oxidation takes place, carbon monoxide being formed, the carbon content of the melt is reduced and silicon and manganese are also oxidized. Finally deoxidizers (ferromanganese, ferrosilicon, aluminium) are added just before tapping, or as the steel is run into the ladle, to improve the quality of the steel. Ferromanganese (80% Mn, 6% C, remainder iron) reacts with the iron oxide to give iron and manganous oxide, which is insoluble in steel, and the excess manganese together with the carbon adjusts the composition of the steel.

The acid process is now declining in use and the basic process produces normal grades of steel. The basic slag, rich in phosphorus, is used as a fertilizer.

Oxygen steel-making

With the introduction of the tonnage oxygen plant situated near the steel furnaces, oxygen is now available at competitive cost in large volumes to greatly speed up steel production. The oxygen plant liquefies air and the oxygen is then fractionally distilled from the nitrogen and argon and is stored in the liquid form as described in the section on liquid oxygen. In the open hearth process an oxygen lance is arranged to blow large volumes of oxygen onto the molten metal in the hearth. With the use of oxygen instead of air there is minimal nitrogen introduced into the steel (below 0.002%), so that its deep-drawing qualities are improved and the time for converting the charge into steel is reduced by as much as 50%. It appears that the basic open hearth process is being replaced by the basic oxygen furnace (BOF) in various forms in Britain, Europe, USA and Japan. The advantages are that iron ores of variable phosphorus content can be used and there are reduced labour and refractory costs. These factors together with the capability of the process to use up to 40% scrap make for reduced costs and higher efficiency.

Fig. 2.3

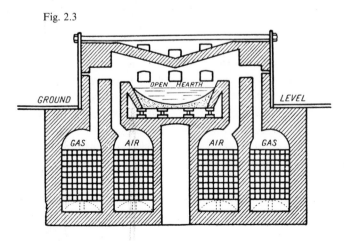

GROUND

OPEN HEARTH

LEVEL

GAS AIR AIR GAS

When the oxygen is blown onto the molten charge it reacts to form iron oxide, which combines with lime present as flux to form an oxidizing slag on the melt. Reactions occur between the molten metal and the slag resulting in the removal of phosphorus, and lowering of the silicon, manganese and carbon content.

Basic oxygen steel (BOS)

The *basic oxygen steel-making* process together with the electric arc process is responsible for much of present-day steel production. A typical basic oxygen furnace consists of a steel-cased converter lined with dolomite holding up to 400 tonnes of metal. Hot metal is transported in torpedo ladles to the hot-metal pouring station equipped with extraction facilities for fume and kish (graphite which separates from and floats on top of the charge), and selected torpedoes are desulphurized with calcium carbide. The hot pig iron, scrap, flux and any alloys are added and oxygen is injected through a multi-holed, water-cooled lance from a near-by tonnage oxygen plant, onto the surface of the charge. Oxidation is rapid, with the blowing time lasting about 17 minutes, and the converter is then tilted and emptied. The BOF takes between 25 and 35% of scrap metal and efforts are being made to increase this percentage because scrap is cheaper than hot pig iron (Fig. 2.4 *a,b,c,d,e*).

The Maxhütte bottom-blown furnace (OBM) is a steel-cased converter lined with dolomite with special tuyères in the base of the furnace through which oxygen and powdered lime are introduced. This in turn is surrounded by a protective shield of hydrocarbon fuel gas (natural gas, propane) to protect the refractory lining (Fig. 2.5). The remainder of the process is similar to that already described and the benefits claimed for this process are absence of fuming and splashing and a lower final carbon and sulphur content in the steel, the injection of lime ensuring rapid removal of the phosphorus.

Electric arc steel-making

When a welder uses a carbon electrode to produce an arc and give a molten pool on a steel plate he is using the same basic principle as that which is done on a very much larger scale in the electric arc furnace. The heat in this type of furnace comes from the arcs struck between three carbon electrodes connected to a three-phase electric supply and the charge in the furnace hearth, and thus no electrode is required in the furnace hearth.

The furnace shell is mild steel lined with refractories, with two doors (one on smaller furnaces). The three electrodes project through the refractory-

Fig. 2.4. Basic oxygen steel-making.
(*a*) Charging with scrap. The converter is tilted and charged with scrap from a charging box which tips the scrap into the previously heated converter. The scrap represents up to 30% of the total charge.

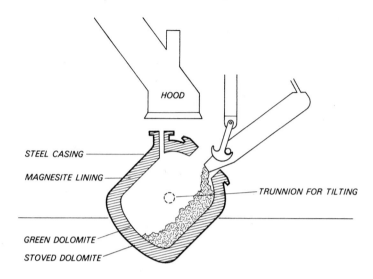

Fig. 2.4 (*b*) Charging with molten iron. Molten iron from the blast furnace is brought to the converter in 'torpedo-cars' and transferred by ladle into the converter.

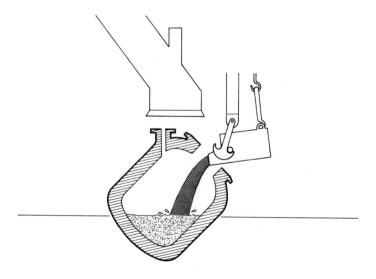

(*c*) The fume-collecting hood is lowered on to the furnace neck. A water-cooled oxygen lance is lowered to within a metre or so of the molten metal surface. Oxygen is blown through the lance and causes turbulence and rise in temperature of the metal. Impurities are oxidized with the 'blow' lasting about fifteen minutes during which time temperatures are carefully controlled and analysis made of the molten metal.

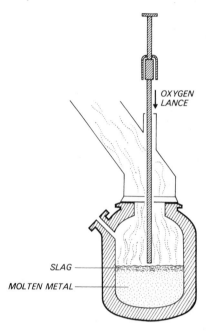

(*d*) When temperature and metal analysis are satisfactory the hood is lifted, the converter tilted and the steel poured from below the slag which has formed into the teeming ladle from which it passes to the continuous casting plant and is cast into ingots.

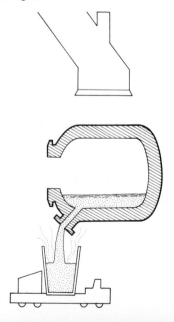

Fig. 2.4. (*e*) Finally the converter is tilted in the opposite direction and the slag which remains is poured into a slag ladle.

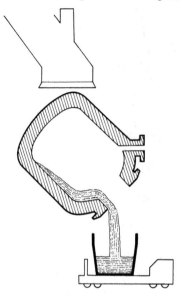

Fig. 2.5. Maxhütte bottom-blown oxygen furnace (OBM).

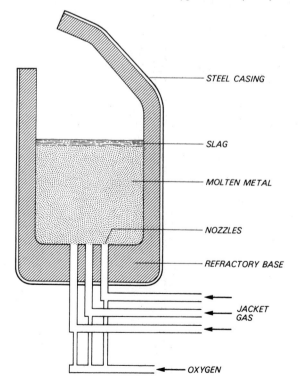

lined roof down to the level of the surface of the charge, and roof and electrodes are arranged to swing aside to clear the furnace shell for charging or repair, this type of furnace being batch-charged, though developments are proceeding to feed furnaces continuously either through the roof or side walls. A sealing ring on top of the side walls supports the weight of the roof and the furnace can tip about 15° towards the main door and about 50° forward for tapping. Some of the larger furnaces can also rotate so as to give a variety of melting positions for the electrodes (Fig. 2.6).

Since it is necessary to remove phosphorus and sulphur a basic slag is required, and the furnace must be lined with basic refractories such as dolomite or magnesite for roof, side walls and hearth to give the basic electric arc process. The acid process is only used in cases where melting only is to be done, with little refining.

Large transformers of the order of 80–100 MVA capacity feed from the grid supply to that for the arcs. The voltage drop across the arc is a function of the arc length, and the greater the voltage drop, the greater the power for a given arc current. As a result the secondary voltage to the arc may be 100–600 V, with currents up to 80 000 A in large furnaces. The three graphite electrodes can vary from 75 to 600 mm in diameter and in length from 1.2 to 2.5 m, and can be raised or lowered either hydraulically or by electric motor, this operation being done automatically so as to keep the arc length correct. Current to the electrodes is taken via water-cooled clamps and bundles of cables to give flexibility.

Fig. 2.6. Electric arc furnace.

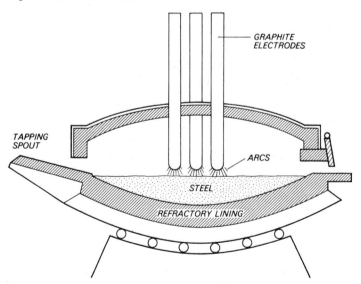

One of the greatest advantages of the arc furnace is that it can deal with up to 100% scrap charge, and whereas in the past it was used for making high-grade alloy steel it is now used to melt high-percentage scrap charges and even to produce ordinary grades of steel.

The charge is of scrap, iron ore, blast furnace iron, DR iron and limestone, depending upon availability. The electrodes are lowered onto the charge, the arcs struck and melting proceeds. The oxygen necessary for the removal of impurities is obtained from the iron ore charge, the furnace atmosphere and in many cases by lance injection of oxygen, the silicon, manganese and phosphorus being removed by oxidation and entering the slag. The carbon is oxidized to carbon monoxide, which burns to carbon dioxide.

Melting is done under a basic oxidizing slag, black in colour, and when the desired level of carbon and phosphorus is obtained, the slag is thoroughly removed and the melt deoxidized with ferrosilicon or aluminium. A reducing slag is now made using lime and anthracite or coke dust and the lime reacts with the iron sulphide to form calcium sulphite and iron oxide; the calcium sulphite is insoluble in steel and thus enters the slag, removing the sulphur. This removal requires reducing conditions in the furnace, which is not possible with any other furnace in which oxygen is used to burn the fuel so that this sulphur removal is another great advantage of the arc furnace. Alloy additions are made under non-oxidizing conditions which give good mixing, and carbon can be added if required (recarburation) in the form of graphite or coke.

Vacuum refining

Further improvement in steel quality is obtained by vacuum refining. An example is Vacuum Oxygen Decarburization (VOD). A stream of oxygen from a lance is blown onto the surface of the molten steel under partial vacuum in a vacuum chamber. Argon is bubbled through the melt for stirring and alloy additions are made from the top. With this method the carbon content of the steel can be reduced to 0.08% (and lower with higher vacuum), hydrogen and nitrogen contents are reduced and chromium addition recoveries are high.

Induction furnaces

These are melting furnaces generally used for the production of special steels in sizes from 100 to 10 000 kg and give accurate control over the steel specification (Fig. 2.7).

When discussing the principle of the transformer we will see (p. 215) that if an alternating current flows in the primary winding, an alternating current is generated in the secondary winding. The alternating magnetic flux due to the primary current generates, or induces, a current in the secondary circuit. In the induction furnace the primary coil is wound around the refractory crucible which contains the metallic charge to be melted. When an alternating current flows in the primary coil eddy currents are induced in the charge and generate the heat required for melting. As the frequency of the alternating current increases, the eddy currents, and thus the heating effect, increase.

Furnaces operate at mains frequency (50 Hz) or at 100, 150, 800, 1600 Hz, etc., and high-frequency furnaces employing static converters to change the frequency operate from 10 to 15 kHz and at voltages up to several kV. The eddy currents produce a stirring action which greatly improves temperature control.

The furnace refractories can be a pre-cast crucible for the smaller furnaces or have a rammed lining of magnesite or, in some cases, silica. The hollow square-shaped copper conductors of the coil are closely wound and water-cooled, this being an essential feature to prevent overheating and consequent breakdown. Small furnaces up to 50 kW are very convenient for laboratory and research work and can have capacities as low as a few kilograms. Clean scrap of known analysis can be used in the charge, and as oxidation losses are minimal and there is little slag, there is practically no loss of alloying elements during the melt.

Fig. 2.7. Typical coreless induction furnace in capacities of 0.75–10 tonnes.

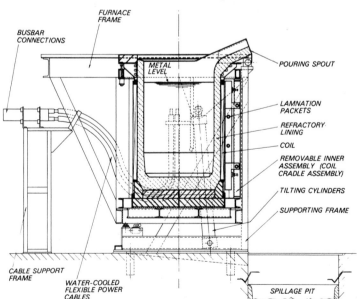

Malleable cast iron

This is made from white cast iron by annealing or graphitizing. The white cast iron is packed in haematite, heated to about 900 °C and kept at this for two or three days, after which the temperature is slowly reduced. In this way, some of the combined carbon of the white cast iron is transformed into free carbon or graphite. Malleable castings are used where strength, ductility and resistance to shock are important, and they can be easily machined.

The 'blackheart' process is similar to the Réaumur or 'whiteheart' process just described except that bone dust, sand and burnt clay are used for packing in place of iron oxide, the temperature being about 850 °C. This converts the combined carbon in the cast iron into temper carbon, and after the treatment they contain little or no combined carbon and about $2\frac{1}{2}\%$ graphite. The castings prepared by the former method show a grey fractured surface with a fine grain like mild steel, while those made by the 'blackheart' process have a black fracture with a distinct white rim.

A typical composition of malleable iron is: carbon 2–3%, silicon 0.6–1.2%, manganese under 0.25%, phosphorus under 0.1%, sulphur 0.5–0.25%.

The effect of the addition of carbon to pure iron

We have seen that the chief difference between iron and mild steel is the amount of carbon present. Steel may contain from 0.03 to 2% carbon, mild or soft steel containing about 0.1% carbon and very hard razor-temper steel 1.7–1.9%.

The composition of steel is therefore complicated by these variations of carbon content, and is rendered even more so by the addition of other elements such as nickel, chromium and manganese to produce alloy steels.

Let us consider the structures present in steels of various carbon contents which have been cooled out slowly to room temperature. If we examine a highly polished specimen of wrought iron under a microscope magnifying about 100 times (× 100), Fig. 2.8a, we can see the white crystals of ferrite with the crystal boundaries and also dark elongated bands, which are particles of slag entrapped during the rolling process. A specimen with no inclusions is shown in Fig. 2.8b. Now examine a specimen of 0.2% carbon steel under the same magnification. It shows dark areas in with the whiter ferrite, Fig. 2.8c. A 0.4% carbon steel appears with more dark areas, Fig. 2.9c, so that it is evident that an increase in carbon content produces an increase in these dark areas, which if observed under first a magnification of × 1000, Fig. 2.9a, and then of × 2500, Fig. 2.9b, are seen to consist of a layered structure, darker areas alternating with ligher ones. The dark areas are iron carbide (Fe_3C) or cementite formed by the chemical combination of ferrite and carbon thus: ferrite + carbon → cementite, or iron + carbon → iron carbide. These alternating layers of ferrite and cementite are called pearlite since they have a mother-of-pearl sheen when illuminated. Pearlite contains 0.85% carbon and is known as a eutectoid.* When a steel contains 0.85% carbon the structure is all pearlite and if more than this percentage is present we find that the carbon has combined with more ferrite reducing the area of pearlite and forming cementite in the structure. Pearlite is a ductile structure while cementite is hard and brittle so that as the carbon content increases above 0.85% and more cementite is formed, the steels become very hard and brittle and steels of more than

* See Chapter 3, eutectoid change.

Fig. 2.8
(*a*) Wrought iron, showing grains of ferrite and slag inclusions. × 100.
(*b*) Ferrite. × 100.
(*c*) 0.2% carbon steel forging normalized, showing pearlite and ferrite. × 100.

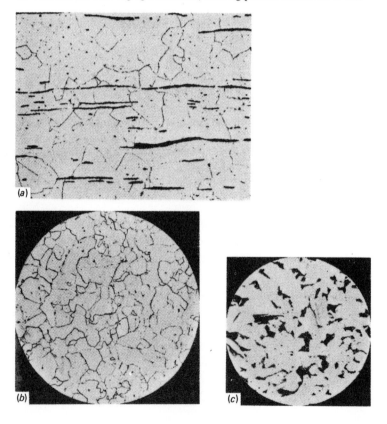

1.7–1.8% carbon are rarely encountered (Fig. 2.9*d* and Fig. 2.10). Above 2% the carbon may be present as free carbon, termed graphite, and when the carbon percentage of the iron is between $2\frac{1}{2}$% and 4% it is known as cast iron. In grey cast iron, which is soft and machinable (but brittle), the carbon is present in the free state as graphite but rapid cooling can cause the carbon to be in the combined form as cementite when we have white cast iron, which is hard and not machinable. Hence in a steel the carbon is always in the combined form while in cast iron it may be present either free as graphite or in the combined form as cementite (Fig. 2.11).

SG cast iron

The flakes of graphite present in grey cast iron which reduce its tensile strength can be changed to sphere-shaped particles by adding to the molten iron small amounts of magnesium (or various other substances). This spheroidal graphite (SG) cast iron has greatly increased strength and

Fig. 2.9
(*a*) 0.8% carbon steel, annealed. × 1000.
(*b*) 0.8% carbon tool steel, annealed. × 2500.
(*c*) 0.4% carbon steel forging, annealed. × 100.
(*d*) Cementite structure in 1.2% carbon tool steel, normalized. × 100.

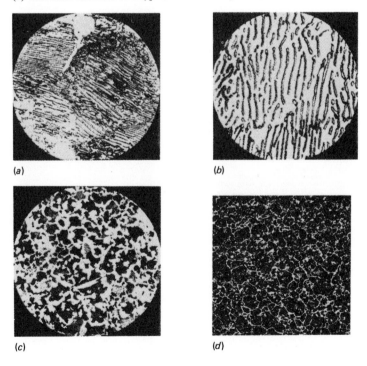

(*a*) (*b*)

(*c*) (*d*)

ductility. In the 'as cast' condition the matrix is normally pearlite, with the carbon in the combined form, and is hard, with a tensile strength about twice that of grey cast iron. Normalizing improves its mechanical properties, and stress-relief heat treatment at 550 °C is recommended for more complicated castings. Annealing the SG iron gives a ferritic matrix, lowering the tensile strength but improving ductility and elongation compared with pearlitic iron. The iron can be hardened by quenching and tempering, and addition of 1–2% nickel increases its hardenability. Nominal composition: C 3.5–3.8%, Si 1.8–2.5%, Mn 0.2–0.6%, P 0.05%, Ni 0–2.0%, Mg 0.04–0.07%.

Carbon and carbon–manganese steels

As the carbon content of a steel increases from 0.05% to about 1.1% the steel changes from a low carbon, soft and malleable steel to a high carbon steel which when heat treated is hard and brittle. Between these values there are a great number of steels suitable for various purposes. All

Fig. 2.10. Structure of steels with varying carbon content.

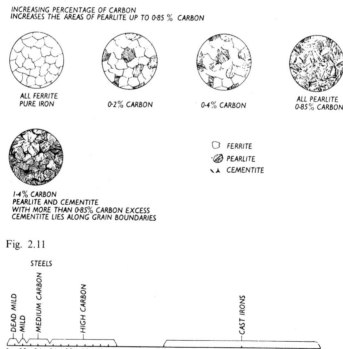

INCREASING PERCENTAGE OF CARBON
INCREASES THE AREAS OF PEARLITE UP TO 0·85 % CARBON

ALL FERRITE
PURE IRON

0·2% CARBON

0·4% CARBON

ALL PEARLITE
0·85% CARBON

1·4% CARBON
PEARLITE AND CEMENTITE
WITH MORE THAN 0·85% CARBON EXCESS
CEMENTITE LIES ALONG GRAIN BOUNDARIES

○ FERRITE
⬤ PEARLITE
↘ CEMENTITE

Fig. 2.11

STEELS

DEAD MILD
MILD
MEDIUM CARBON
HIGH CARBON

CAST IRONS

0   0·2 0·4   0·6   0·8   1·0   1·2   1·4   2·0   4·0

CARBON IN COMBINED FORM
CEMENTITE

CARBON EITHER IN FREE FORM AS GRAPHITE
OR IN COMBINED FORM AS CEMENTITE

Examples: carbon and carbon–manganese steel. BS 970 Part 1

Steel specification	Nearest AISI/SAE Eqvt.	Type	Application	C%	Mn%
030A04	1006	low carbon	pressings and screws	0.08	0.2–0.4
040A12		low carbon	cold forming steel	0.10–0.15	0.3–0.5
040A17		low carbon	cold forming steel	0.15–0.20	0.3–0.5
070M20	1021/1022	20 carbon	general purpose mild steel	0.16–0.24	0.5–0.9
080A30	1030	30 carbon	pipes, valves, flanges	0.28–0.33	0.7–0.9
080A35	1034/1035	35 carbon	high tensile shafts and forgings	0.33–0.38	0.7–0.9
070M55	1055	55 carbon	crankshafts	0.50–0.60	0.5–0.9
060A62	1059	62 carbon	induction hardened gears	0.60–0.65	0.5–0.7
080A72	1070	72 carbon	spring steel	0.70–0.75	0.7–0.9
120M19		19 carbon 1.2 manganese	welded structures	0.15–0.23	1.0–1.4
150M28		28 carbon 1.5 manganese	welded structures	0.24–0.32	1.3–1.7

In the above, sulphur and phosphorous below 0.050%.
AISI = American Iron and Steel Institute, SAE = Society of Automotive Engineers.

steels contain some manganese from the deoxidation process and below 1% it is not considered as an alloying element. If, however, the percentage is increased to about 1.5 we have a range of carbon–manganese steels which have increased tensile strength and are suitable for welded structures.

The designation of these steels (BS 970 Part 1) uses six digits, the first three being from 000 to 299. The plain carbon and carbon–manganese steels fall in the range 000 to 199 and these figures represent 100 times a mean manganese content. Three letters follow these digits, the steel being supplied to: A, analysis; H, hardenability requirements; and M, mechanical properties. The fifth and sixth digits represent 100 times the carbon content of the steel thus:

Example

040A12 manganese content 0.3–0.5%   A, supplied to analysis.
        carbon content 0.1–0.15%   This is a cold forming steel.

The free cutting steels fall within the range 200–240 and the second and third digits indicate 100 times the minimum or mean sulphur content.

The table (p. 81) gives examples of some of the steels and for these the maximum sulphur and phosphorus content is 0.05% in each case. Phosphorus makes the steels cold short (liable to cracking when cold worked). Sulphur makes the steels hot short (liable to cracking when hot worked) hence the very low maximum allowance. The increased sulphur content in the free cutting steels (sulphur up to 0.6% maximum and phosphorus up to 0.07% maximum) improves the machineability but they are difficult to weld and many are not satisfactorily welded.

Alloy steels

An alloy steel may be defined as a steel which owes its properties to the presence of elements other than carbon, manganese up to 1.1% and silicon up to 0.5%.

The purpose of the alloying elements is to give the steel a distinct property which in every case is to increase its toughness, hardness or tensile strength, and to give cleaner and more wear-resistant castings.

Low alloy steels include: (1) structural steels; (2) creep-resistant steels; and (3) high tensile low-alloy steels.

Pre-heating to 150–200 °C according to type is usually required when welding these last steels.

The chief elements which are alloyed with steel are nickel, chromium, manganese (above 1.1% molybdenum, tungsten, vanadium, copper and silicon. We shall only consider very briefly the effect produced on the steel by the alloying elements and not attempt to discuss the variety of steels available under each heading.

Nickel steels

The addition of nickel to a steel increases the strength and toughness. The nickel lowers the critical cooling rate and tends to decompose carbides present to form graphite so that plain nickel steels usually have a lower carbon content. For the higher carbon content nickel steels, manganese, which stabilizes the carbides, is generally added. Nickel tends to form austenite, refines the grain and limits grain growth and gives a range of steels suitable for highly stressed parts such as crankshafts, and axle shafts. The addition of 36% nickel gives a steel with a very low coefficient of thermal expansion.

Examples of direct hardening steels containing nickel. BS 970 Part 2

Steel specification	Application	C %	Si %	Mn %	S%	P%
503M40	crankshafts	0.36–0.44	0.10–0.35	0.70–1.00	0–0.5	0–0.40
503H37		0.33–0.41	0.10–0.35	0.65–1.05	0–0.50	0–0.40

Chromium steels and nickel-chromium (stainless) steels

Chromium has the opposite effect from nickel on steel because it raises the upper critical temperature, tends to form ferrite, and, due to the formation of carbides, it increases the hardness and strength but reduces the ductility and promotes grain growth. Overheating and maintaining at elevated temperature for too long a period should thus be avoided.

Examples of direct hardening steels containing chromium. BS 970 Part 2

Steel specification	Applications	C %	Si %	Mn %	Cr %	S %	P %
530A30	gears and connecting rods	0.28–0.33	0.10–0.35	0.60–0.80	0.90–1.20	0–0.50	0–0.40
530/H36		0.33–0.40	0.10–0.35	0.50–0.90	0.80–1.25	0–0.50	0–0.40

When greater amounts of chromium are added the steel becomes resistant to corrosion, a thin layer of resistant chromium oxide forming on the surface.

If the steel contains about 11% chromium it may be termed stainless and it is convenient to consider two classes: A, ferritic and martensitic; and B, austenitic, according to which phase is predominant, the austenitic group being alloyed with nickel and chromium.

Group A: Martensitic. These steels are alloyed mainly with chromium, 11–18%, and harden upon cooling from welding temperatures, resulting in embrittlement and a tendency to crack in the weld and heat-affected zone so that in general they are not recommended for welding. These steels are hardenable by heat treatment, are magnetic and have a lower coefficient of expansion and a lower thermal conductivity than mild steel. They are used in engineering plant subject to mildly corrosive conditions, for cutlery, sharp-edged instruments, ball and roller bearings, etc., according to the carbon and chromium content.

If welding is to be carried out the work should be pre-heated to 200–400 °C, followed by slow cooling after welding to reduce the hardness and danger of cracking. Post-heating to 650–700 °C is also advised and a basic electrode of the 19% Cr, 9% Ni, 3% Mo type is used. Electrodes of similar composition to the parent metal are used only for limited applications such as overlaying.

Examples of this class of steels are:

C%	Cr%	Si%	Mn%	Other elements	Weldability
0.28–0.36	12–14	0.8	1.0	—	Not generally recommended.
0.12	11.5–13.5	0.8	1.0	—	Poor brittle welds but type with 0.06% C max., fair.
0.7–0.9	15.5–17.5	0.8	1.0	0.3–0.7 Mo	Poor. Should not be welded.

Group A: Ferritic. These steels are alloyed with chromium and because of their low carbon content the structure is almost completely ferritic. They are magnetic and though easier to weld than the martensitic group because they are not hardenable to any extent by heat treatment, they suffer from grain growth and embrittlement at temperatures above 900 °C, and from a form of intergranular corrosion in the HAZ.

When welding these steels pre-heating to 200 °C is recommended followed by post-heat at 750 °C, which helps to restore ductility. An austenitic stainless steel electrode is recommended for mildly corrosive conditions but if the application is in sulphur-bearing atmospheres these attack the nickel. An example of this ferritic type is: C 0.08%, Cr 16–18%, Si 0.8%, Mn 1.0%. Weldability is fair, but welds tend to be brittle.

Group B: Austenitic. These steels are alloyed with chromium and nickel. The presence of nickel makes the steel austenitic at room temperature, confers high temperature strength, helps corrosion resistance and controls the grain growth associated with the addition of chromium. The chromium tends to form carbides while the nickel tends to decompose them, so that in this group of steels the disadvantages of each alloying element are reduced. The addition of molybdenum to these steels is to improve corrosion resistance and high-temperature strength. (Temper brittleness may occur when the steels are tempered in the range 250–400 °C.) It is the high-nickel, high-chromium steels which are of great importance in welding since much fabrication is done in these steels.

Those steels containing 17.5–19.5% chromium and 8–10% nickel, known as 18/8 from their nearness to this composition, harden with cold work, are non-magnetic, resistant to corrosion, can be polished, machined and cold-worked. They have a coefficient of thermal expansion 50% greater than mild steel but the thermal conductivity is much less than mild steel so that there is a narrower HAZ when they are welded. They are used in chemical, food, textile and other industrial plant subject to corrosive attack, also for domestic appliances and decorative applications.

Weld decay

When the austenitic steels are heated within the range 600–850 °C, carbon is absorbed by the chromium and chromium carbide is precipitated along the grain boundaries (Fig. 2.12). As a result, the chromium content of the austenite in the adjacent areas is reduced and hence the resistance to corrosive attack is lowered. When welding, a zone of this temperature range exists near the weld and runs parallel to it, and it is in this zone that the corrosion may occur and is known as weld decay, though no corrosive effect occurs in the weld itself. Heat treatment, consisting of heating the part to 1100 °C and water quenching, restores the carbon to solid solution but has the great drawback that much of the fabricated work is too large for heat treatment.

The difficulty is overcome by adding small quantities of titanium, niobium (columbium) or molybdenum to the steel. These elements form carbides very easily and thus no carbon is available for the chromium to form carbides. The austenitic steels with these additions are known as stabilized steels and they contain a very low percentage of carbon (0.03–0.1%). They have good welding properties and need no subsequent heat treatment. The steels with 0.03% carbon may have no stabilizers added but are suitable for welding because the low carbon content precludes the formation of carbides.

Electrodes of 18% Cr–8% Ni; 19% Cr–10% Ni; and 25% Cr–12% Ni, with or without Mo, Nb, Ti and W produce welds which contain residual ferrite and are resistant to hot cracking when the welds are under restraint. Grades with a higher ferrite content under certain conditions in the temperature range 450–900 °C may lose ductility and impact resistance due to the transformation of ferrite into the brittle sigma phase. Most stainless steels however do not encounter these conditions and thus embrittlement does

Fig. 2.12

(*a*) Carbide precipitation at the grain boundaries in an 18/8 class stainless steel. The steel is in a state of heat treatment (500–900 °C) in which it is susceptible to intercrystalline corrosion, but it has not yet been subjected to a severe corrosive medium. × 200.

(*b*) Occurrence of intercrystalline corrosion in an 18/8 stainless steel strip. The steel is in the same condition as (*a*), but it has now been subjected to a severe corroding medium. × 50.

(*c*) The effects of intercrystalline corrosion (weld decay) in an 18/8 steel.

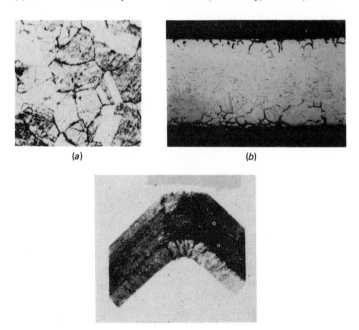

(a)   (b)

(c)

not occur. Manufacturers often indicate the ferrite percentage, e.g. 19% Cr, 9% Ni, 0.05–0.08% C, Nb 10 × carbon content, ferrite 6%.

When welding stainless steel to mild or low-alloy steel, dilution (pp. 107–9) occurs and the weld may suffer 20–50% dilution. Root runs in butt joints are greatly affected since the weld metal is in contact with parent metal on both sides. Additional runs are partly in contact with weld metal already laid down and so suffer less dilution.

If a mild or low-alloy steel electrode is used for welding stainless steel to mild or low-alloy steel the weld metal will pick up about 5% Cr and 4% Ni from the stainless steel plate, resulting in a hardenable, crack-sensitive weld.

An austenitic steel electrode should be selected such that the weld metal will contain not less than 17% Cr and 7% Ni, otherwise there may not be enough ferrite present to prevent subsequent cracking. Electrodes of 20% Cr, 9% Ni, 3% Mo; or 23% Cr, 12% Ni are the most suitable, since their composition ensures that they accommodate the effects of dilution and there is a sufficiently high ferrite content to give resistance to hot cracking. Electrodes of 20% Cr, 20% Ni are also suitable, except for conditions of high restraint.

If mild steel fittings are to be welded to the exterior of stainless vessels, stainless pads can be first welded to the vessel and the fittings welded to the pads. This reduces the danger of penetration of diluted metal to the face subject to corrosive conditions.

BS 2926 covers the range of chromium–nickel austenitic steels and the chromium steels. The code of composition is : first figure or figures is the chromium content, the second the nickel content and the third the molybdenum content. Nb indicates niobium stabilized, L is the low carbon type and W indicates the presence of tungsten. R is rutile coating (usually a.c. or d.c.) and B is a basic coating generally d.c. only, electrode +ve.

Example

19.12.3.L.R. is a 19% Cr, 12% Ni, 3% Mo low carbon (0.03%) rutile-coated electrode.

Identification of stainless and low-alloy steel

If a spot of 30% commercial concentrated nitric acid is placed on grease-free stainless steel, there is no reaction, but on low-alloy or plain carbon steel there is a bubbling reaction.

To determine to which group of stainless steel a specimen belongs, a hand magnet can be used. Ferritic and martensitic steels are stronly magnetic, whilst austenitic steels are generally non-magnetic. However the austenitic steels become somewhat magnetic when cold worked, but there is considerable difference in magnetic properties between them and the ferritic and martensitic types, which is easily detected. The identification of the various grades within the groups is not easy without laboratory facilities, but an indication is given by the hardness after heat treatment, which varies from 400 HV for a 0.12% C type to 700 HV for a 0.7–0.9% C type.

Students wishing to study further the metallurgical aspects of stainless steels should consult the Shaeffler's diagram in Volume 2. In this the various alloying elements are expressed in terms of nickel, which is austenite-forming, and chromium, which tends to form ferrite. Using these values in conjunction with the diagram, information can be obtained on the possible behaviour of the steel during welding.

Steel containing molybdenum

Addition of 0.15–0.3% molybdenum to low-nickel and low-chrome steels reduces the tendency to temper brittleness and gives high impact strength. It increases the strength and creep resistance at elevated temperatures and increases the resistance to corrosion of stainless steels. The nickel–chrome–molybdenum steels have high strength combined with good ductility and are used for all applications in engine components such as shafts and gears involving high stress.

Steel containing vanadium

Vanadium reduces grain growth and, due to the formation of its carbide which can be taken into solid solution, gives strength and resistance to fatigue at elevated temperatures. It is usually used in conjunction with niobium and chromium and in many ways these steels resemble the nickel-chromium steels.

C %	Mn %	Si %	Cr %	Mo %	V %	
0.45	0.6	0.25	1.25	—	0.15 min.	Spring steel.
0.13	1.4	0.25	0.6	0.3	0.1	

Manganese steel

Manganese, in the form of ferromanganese, is used for deoxidation of steel and in most steel there is a manganese content usually less than 1%. Above this value it may be considered an alloying element and lowers the critical temperature. the most widely used manganese steel is that containing 12–14% Mn and 1.2%C. This is austenitic but it hardens greatly with cold work and is widely used for components subject to wear and abrasion since the core retains its toughness while the surface layers

Types of stainless and heat resistant steels. (In each case S max 0.03%, P max 0.04%.)

Steel type	Typical composition %	Applications	Standard product forms
302 (austenitic)	C 0.55, Cr 18.4, Ni 8.9	Water tubing, sinks, exhaust parts, trim, etc.	Sheet, coil, precision strip
304 (austenitic)	C 0.05, Cr 18.4, Ni 9.5	Food and dairy processing equipment, catering equipment, hollow-ware, sinks, etc.	Sheet, coil, precision strip, plate
304L (austenitic)	C 0.02, Cr 18.4, Ni 9.3	Fabricated components for road tankers and process plant, storage tanks	Sheet, coil, precision strip, plate
316 (austenitic)	C 0.55, Cr 17.0, Ni 11.9, Mo 2.5	Process plant parts, pulp and paper equipment, architectural sections	Sheet, coil, precision strip, plate
316L (austenitic)	C 0.02, Cr 17.0, Ni 11.9, Mo 2.5	Process plant parts especially in thick sections	Sheet, coil, precision strip, plate
321 (austenitic)	C 0.05, Cr 17.6, Ni 9.4, Ti 0.4	Heater element tubes, aircraft parts, process plant parts, furnace parts	Sheet, coil, precision strip, plate
347 (austenitic)	C 0.05, Cr 17.6, Ni 9.4, Nb 0.7	Process plant vessels and tubing, aircraft parts	Plate, precision strip
405 (ferritic)	C 0.05, Cr 13.0, Al 0.2	Chemical plant parts, petroleum cracking installations	Plate, precision strip
430 (ferritic)	C 0.55, Cr 16.5	Decorative household and vehicle trim, flatware, interior architectural sections	Sheet, coil, precision strip, plate
410 (martensitic heat treatable)	C 0.12, Cr 12.0	General engineering components pump and gas turbine parts	Precision strip
310 (austenitic)	C 0.55, Cr 24.5, Ni 20.0	Furnace parts, annealing covers heat exchangers, electrical parts	Plate

harden with cold work to an intense degree. These steels have a wide range of applications in earth-moving equipment, rolls, dredger bucket lips and in all cases where resistance to wear and abrasion is of paramount importance.

Steel containing tungsten

Tungsten reduces the tendency to grain growth, raises the upper critical temperature, forms very hard, stable carbides which remain in solution after oil quenching and renders the steel very hard and suitable for cutting tools and gauges.

High-speed steel usually contains tungsten, chromium, vanadium, molybdenum and cobalt with a carbon content of 0.6–1.5%. Tungsten carbide, made by the sintering process, is used for tool tips for cutting tools, is extremely hard and brittle, and is brazed onto a carbon or alloy steel shank.

Nitralloy steels

These steels contain silicon, manganese, nickel, chromium, molybdenum and aluminium in varying proportion, and their carbon content varies from 0.2 to 0.55%. They are eminently suited for purposes where great resistance to wear is required. After being hardened, by the process of nitriding or nitrogen hardening them, they have an intensely glass hard surface and are suitable in this state for crankshafts, camshafts, pump spindles, shackle bolts, etc. (see Heat Treatment).

Carbon equivalent

As the carbon content and the amount of alloying elements increases, care has to be taken, when welding these steels, of their tendency to crack. As there is a great variety of these steels it is convenient to express the carbon content and the percentage of each alloying element in terms of the 'carbon equivalent' of the steel. In this way the alloy steel is expressed in terms of a 'carbon content' and the formula used is BS 4360. The CE is thus an indication of the tendency to crack when welded.

$$CE\% = C\% + \frac{Mn\%}{6} + \frac{Cr\% + Mo\% + V\%}{5} + \frac{Ni\% + Cu\%}{15}$$

For weldable structural steels the CE is specified and the electrode manufacturers charts should be studied as to which electrode and which type of coating (basic or rutile) is suitable. In general, for high-tensile requirements with higher impact properties basic coated electrodes are advised. (See Vol. 2, pp. 76–7 for a worked example.)

The effect of heat on the structure of steel

Suppose we heat a piece of steel containing a small percentage of carbon and measure its temperature rise. We find that after a certain time, although we continue supplying heat to the steel, the temperature ceases to rise for a short time and then begins to rise again at a uniform rate. Evidently at this arrest point, termed a critical point, the heat which is being absorbed (decalescence), since it has not caused a rise in temperature, has caused a change to occur in the internal structure of the steel.

If the heating is continued, we find that a second arrest or critical point occurs, but the effect is not nearly as marked as the first point. At a higher temperature still, a third critical point occurs, similar in effect to the first.

If the steel is now allowed to cool at a uniform rate, we again have the three critical points corresponding to the three when the steel was heated, but they occur in each case at a slightly lower temperature than the corresponding point in the heating operation. At these points in the cooling operation, the metal gives out heat but the temperature remains steady. The evolution of heat on cooling through the critical range and which is visible in a darkened room is known as recalescence. The second arrest points which occur between upper and lower critical temperatures involve the loss and gain of magnetic properties and need not concern us here.

If the experiment is done with steels of varying carbon content it will be found that the lower critical point is constant at about 720 °C for all steels, but the temperature of the upper critical point decreases with increasing carbon content, until at 0.85% carbon the two critical points occur at the same temperature. Figure 2.13 is part of the iron–carbon equilibrium or

Fig. 2.13

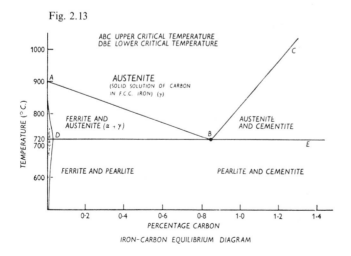

IRON–CARBON EQUILIBRIUM DIAGRAM

constitutional diagram which shows how the structure of any plain carbon steel changes with temperature and shows the limits of temperature and composition in which the various constituents are stable (see Chapter 3 on equilibrium diagrams). We have seen that iron below 900 °C has a body-centred cubic structure (α iron), and in this state it can only have a trace of carbon in solution (up to 0.03% at 700 °C). When, however, the temperature rises above its upper critical point and the structure changes to face-centred cubic (γ iron), this type of iron can have up to 1.7% carbon in solution before it is saturated. Thus when a steel is heated above its upper critical point the carbon, which was present in the combined form as cementite, dissolves in the iron to form a solid solution of iron and carbon, that is, dissolves completely in the iron although the iron is still in the solid state. This solid solution is called austenite (see Fig. 2.15*a*) and the carbon is diffused uniformly throughout the iron. Now let us lower the temperature of a 0.3% carbon steel slowly from above the upper critical point, about 850 °C. At the upper critical point the structure begins to change. Body-centred cubic crystals of ferrite are precipitated and the carbon content of the remaining austenite begins to increase. This continues as the temperature falls until at the lower critical point the austenite contains 0.85% carbon; it is saturated. The austenite now precipitates cementite (ferrite and carbon chemically combined) at the lower critical point, and it does this in alternate layers with the ferrite that is separating out, forming the areas of pearlite (Fig. 2.14). A 0.5% carbon steel will change in the same way, but

Fig. 2.14

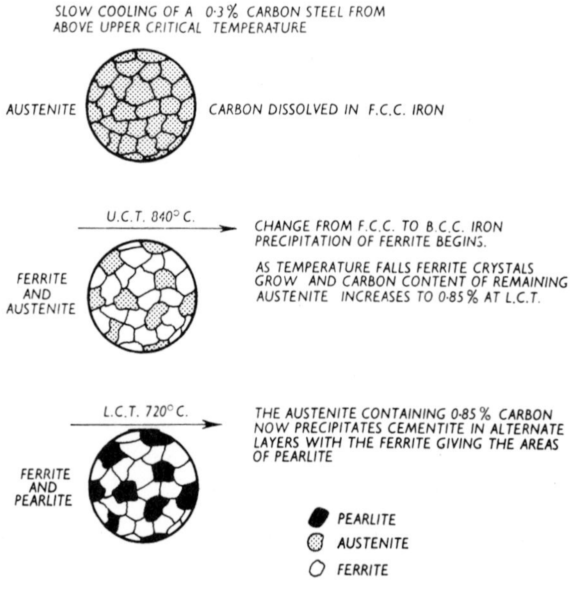

SLOW COOLING OF A 0.3% CARBON STEEL FROM ABOVE UPPER CRITICAL TEMPERATURE

AUSTENITE   CARBON DISSOLVED IN F.C.C. IRON

U.C.T. 840° C.

CHANGE FROM F.C.C. TO B.C.C. IRON PRECIPITATION OF FERRITE BEGINS.

AS TEMPERATURE FALLS FERRITE CRYSTALS GROW AND CARBON CONTENT OF REMAINING AUSTENITE INCREASES TO 0.85% AT L.C.T.

FERRITE AND AUSTENITE

L.C.T. 720° C.

THE AUSTENITE CONTAINING 0.85% CARBON NOW PRECIPITATES CEMENTITE IN ALTERNATE LAYERS WITH THE FERRITE GIVING THE AREAS OF PEARLITE

FERRITE AND PEARLITE

● PEARLITE
◉ AUSTENITE
○ FERRITE

begins the change at a lower temperature since the upper critical point is about 800 °C. In the case of a 0.85% carbon steel the transformation begins and ends at about 720 °C, since upper and lower critical points are at the same temperature, and the final structure will be all pearlite. A steel with more than 0.85% carbon will begin to precipitate carbon in the form of cementite at the upper critical point since carbon is in excess of 0.85%. At the lower critical point this transforms to pearlite so that the final structure is pearlite and cementite.

Heat treatment of steel

Hardening. In order to harden a carbon steel it must be heated to a temperature of 20–30 °C above its upper critical temperature and kept at this temperature long enough to ensure that the whole mass is at this

Fig. 2.15
(*a*) Austenite in an 18% chromium 8% nickel steel sensitized and etched to show the grain boundaries. × 500.
(*b*) Martensite. × 250.
(*c*) Bainite in a low chromium nickel and molybdenum steel transformed over the temperature range 570–430 °C. × 500.

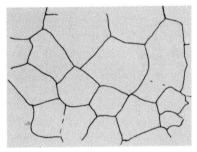

(a)

(b)

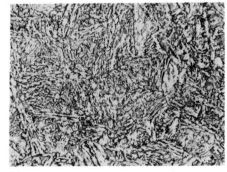

(c)

temperature and the structure is austenitic. Maximum hardness can now be obtained by quenching the steel in water or brine.

Examined under the microscope the structure appears as fine needle-like (acicular) crystals and is known as martensite (Fig. 2.15*b*); and the steel is hard and brittle. The rapid quench has prevented the normal change from austenite to ferrite and pearlite taking place. Quenching less severely produces bainite, a structure which when viewed under high magnification can be seen as an aggregate of ferrite and carbide particles like finely divided pearlite (Fig. 2.15*c*). Martensite and bainite are often found together in quenched steels.

The rate of cooling is measured by the fall in temperature per second and can vary from a few to some hundreds of degrees depending upon the method used. The Critical Cooling Rate for a steel is the lowest rate at which a steel can be quenched to give an all-martensitic structure. At lower rates bainite and/or finely divided pearlite will form. If the steel has a large mass the outer layers will cool quickly when quenched, giving maximum hardness, while the core will cool much more slowly and will be softer (mass effect). Although simple shapes quench successfully, more complicated shapes may suffer distortion or cracking, or internal stresses may remain. In this case where a rapid quench would lead to complications, an alloy steel (e.g. one containing nickel) can be used instead of a plain carbon steel. The nickel lowers the critical cooling rate by slowing up the rate of transformation of austenite into its decomposition products so that a less drastic quench is required to produce martensite, reducing the mass effect and the risk of distortion, cracking and internal stresses.

Air-hardened steels usually contain sufficient additions of nickel and chromium to reduce the critical cooling rate so that the steel is hardened by cooling in air, followed by tempering as required.

Quenching media include the following in decreasing order of quenching speed: caustic soda (sodium hydroxide), 5% solution; brine, 5–25% solution; cold water; hot water; mineral, animal, vegetable oils.

Tempering. The hardness and brittleness of a rapidly quenched steel together with the possibility that there may be internal stresses in the steel make the steel unsuitable for use unless the greatest possible hardness is required. The hardness and brittleness can be reduced and internal stresses relieved by tempering, in which process the hard and brittle martensitic structure is transformed to softer and more ductile structures but yet harder and tougher than ferrite and pearlite. To temper a steel it is re-heated to a definite temperature after hardening and then cooled.

(*a*) Heating to the range 200–250 °C relieves immediate lattice stress but the overall stress pattern persists to fairly high temperatures.

(*b*) Diffusion of carbon from martensite begins at about 150 °C and is practically complete at 350 °C. Heating to the range 150–350 °C forms a mixture of finely divided ferrite and cementite, not so hard but tougher than martensite.

(*c*) Coalescence of the carbides (cementite) starts at about 350 °C and is almost complete by about 650 °C, thus tempering in this range produces a structure similar to that in (*b*) but with larger particles and is softer and more ductile, the particle size depending upon the precise tempering temperature (Fig. 2.16*b*).

The structures obtained by tempering are stages in the austenite–pearlite transformation due to variations in the size and shape of the carbide particles and the way in which they are found in the ferrite matrix.

Interrupted quenching processes reduce internal stresses and distortion and reduce the possibility of quench cracking. If a steel is heated above its upper critical temperature and is then quenched in a bath of molten metal (lead or tin) or salt, kept at a fixed temperature, the quench is not so drastic and there is less temperature gradient between surface and core, reducing

Fig. 2.16
(*a*) Spheroidised carbides in a 1% carbon, 1% chromium steel. Annealed. × 700.
(*b*) Variation in particle size due to different tempering temperatures in a low alloy steel. × 250.

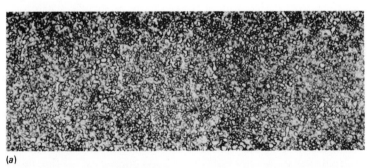

(a)

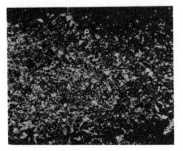 

(b)

stress and distortion effects. If the steel is held at this temperature for varying periods of time and is then quenched out, time–temperature-transformation curves can be drawn indicating the various structures obtained by varying time and temperature.

A process termed martempering can be used to obtain a martensitic structure without the disadvantages of a drastic quench. The steel is quenched from the austenitic condition into a bath of molten metal kept at a temperature just above that at which martensite can form (260–370 °C) until it has a uniform temperature throughout and is then cooled in air, the structure being fine-grained martensite, and the thermal stresses are minimized. Austempering is an interrupted quenching process in which the steel is quenched from the austenitic condition into a bath of molten metal kept at a temperature below the critical range but above the temperature at which martensite can form. It is held at this temperature until complete transformation has occurred and then cooled to room temperature, the structure being pearlite and bainite.

A method often used to obtain a temper on a cutting tool consists of raising the part to bright red heat and then quenching the cutting end of the tool in water. The tool is then removed, any oxide that has formed is quickly polished off and the heat from the part which was not quenched travels by conduction to the quenched end, and the temper colours, formed by light interference on the different thicknesses of layers of surface oxide, begin to appear. The tool is then entirely quenched, when the required colour appears. The colours vary from pale yellow (220 °C), through straw, yellow, purple brown, purple, blue to dark blue (300 °C).

Accurate control of tempering temperature can be obtained by using: furnaces with circulating atmospheres; oils for the lower temperature range; liquid salt (potassium nitrate, sodium nitrite, etc.); or liquid metal, e.g. lead.

Annealing. Annealing is the process by which the steel is softened, and internal strains are removed. The process consists of heating the metal to a certain temperature and then allowing it to cool very slowly out of contact with the air to prevent oxidation of the surface. After the first softening is obtained, if the annealing is prolonged, large crystals are formed. These, as is usual with all crystals, grow in size and decrease in number as the annealing continues. This is known as crystal growth or grain growth (Fig. 2.17).

As these grow in size, the resistance of the metal to shock and fatigue is greatly lowered; hence over-annealing has the bad effect of promoting grain growth, resulting in reduction in resistance to shock and fatigue.

The annealing temperature should be about 50 °C above the upper critical point and therefore varies with the carbon content of the steel. Low-carbon steel should therefore be heated to about 900 °C, while high-carbon steel should be heated to about 760 °C.

Use is made of the fact that iron (ferrite) recrystallizes at 500–550 °C in the treatment known as process annealing, which is used for mild steel articles which have been cold worked during manufacture. They can be packed in a box with cast iron filings over them, the lid luted on with clay, and then heated to 550–650 °C. Recrystallization of the ferrite takes place (there is little pearlite in the structure) and they are allowed to cool out very slowly in the box, after which considerable softening has taken place (Fig. 2.18).

High-carbon steels whose structure is mostly pearlite can be annealed by heating to 650–700 °C. At this temperature the cementite forms or balls up into rounded shapes, and the steel is softer and may be drawn and worked. At temperatures above this, pearlite is reformed and the steel becomes hard (Fig. 2.16a).

Fig. 2.17. Grain growth due to normalizing at increasingly high temperatures. × 100.
(a) 0.4% carbon steel normalized at 850 °C.
(b) 0.4% carbon steel normalized at 1000 °C.
(c) 0.4% carbon steel normalized at 1200 °C.

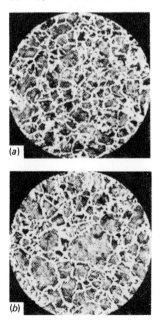

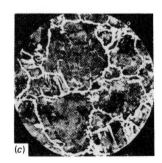

Normalizing. This process consists in raising the steel only slightly above the upper critical point, keeping it at this temperature for just sufficient time to heat it right through, and then allowing it to cool as rapidly as possible in air. This causes a refining of the structure, since recrystallization takes place, and a coarse structure becomes much finer, since the steel is not held at the high temperature long enough for any grain (or crystal) growth to take place (see Fig. 2.17).

Overheated steel. If steel is exposed to too high a temperature or for too long a time to temperatures above the upper critical point it becomes overheated. This means that a very coarse structure occurs and, on cooling, this gives a similar coarse structure of ferrite and pearlite. This structure results in great reduction in fatigue resistance, impact strength, and a reduced yield point, and is therefore undesirable.

Steel which has been overheated is therefore extremely unsatisfactory. Correct heat treatment will restore the correct structure.

Burnt steel. If steel is heated to too high a temperature, this may result in a condition which cannot be remedied by heat treatment and the steel is said to be 'burnt'. This condition is due to the fact that the boundaries of the crystals become oxidized, due to absorption of oxygen at high temperature, and hence the steel is weakened (Fig. 2.19).

Case-hardening. Case-hardening (and also pack-hardening) is a method by which soft low-carbon steel is hardened on the surface by heating it in contact with carbonaceous material (material containing carbon). Parts to be case-hardened are packed in a box and covered with carbonaceous powder, such as charred leather, powdered bone, animal charcoal, or

Fig. 2.18. Heat treatment of steel.

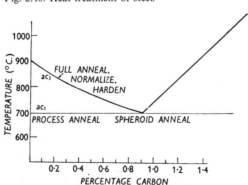

cyanide of potassium (KCN). The box is then placed in the furnace and heated above the critical temperature (that is, above 900 °C, depending on the carbon percentage in the steel). The steel begins to absorb carbon at red heat and continues to do so, the carbon diffusing through the surface. The box is then removed from the furnace and the parts on being taken out can either be directly water or oil quenched. Another favoured method is to allow them to cool out slowly, then heat up to about 800 °C, and quench in oil or water, depending on the hardness required in the case.

In the process known as gas carburizing, carbon is introduced into the surface of the part to be hardened by heating in a current of a gas with a high carbon content, such as hydrocarbons or hydrocarbons and carbon monoxide. This process is extensively used today and lends itself to automation with accurate control and uniformity of hardness.

The case-hardening furnace is almost always found nowadays as part of the equipment of large engineering shops. Parts such as gudgeon pins, shackle bolts and camshafts and, in fact, all types of components subject to hard wear are case-hardened.

The drawback to the process is that, owing to the quenching process, parts of complicated shape cannot be case-hardened owing to the risk of distortion.

The percentage of carbon in steels suited to case-hardening varies from 0.15 to 0.25%. Above this, the core tends to become hard. The carbon content of the case after hardening may be as high as 1.1%, but is normally about 0.9% to a depth of 0.1 mm.

Nitriding or nitrarding. This process consists of hardening the surface of 'nitralloy steels' (alloy steels containing aluminium and nickel) by heating the steel to approximately 500 °C in an atmosphere of nitrogen. The steel to

Fig. 2.19. Oxidation along crystal boundaries in mild steel which has been overheated and 'burnt'.

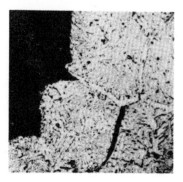

be nitrarded is placed in the furnace and ammonia gas (NH_3) is passed through it. The ammonia gas splits up, or 'cracks', and the nitrogen is absorbed by the steel, while the hydrogen combines with the oxygen and steam is formed, passing out of the furnace. The parts are left in the furnace for a period depending on the depth of hardening required, because this process produces a hardening effect which decreases gradually from the surface to the core and is not a 'case' or surface hardening. When removed from the furnace, the parts are simply allowed to cool. The nitralloy steel is annealed before being placed in the furnace and the parts can be finished to the finest limits, since the heat of the furnace is so low (500 °C) that distortion is reduced to a minimum, and there is no quenching. Nitralloy steel, after the nitrarding process, is intensely hard, and it does not suffer from the liability of the surface to 'flake' as does very hard case-hardened steel.

It is used extensively today in the automobile and aircraft industries for parts such as crankshafts, pump spindles, etc.

The effect of welding on the structure of steel

A typical analysis of all-weld metal-deposit mild steel is: carbon 0.06–0.08%, manganese 0.43–0.6%, silicon 0.12–0.4%, sulphur 0.02% max., phosphorus 0.03% max., remainder iron. During the welding process, the molten metal is at a temperature of from 2500 to 3000 °C and the weld may be considered as a region of cast steel. Since regions near the weld may be comparatively cool, giving a steep thermal gradient from weld to parent plate, it will be possible to find crystal structures of all types in the vicinity, and great changes may take place as the rate of cooling is altered.

A typical cross section from the molten pool to the cold section of the parent plate might reveal the following regions (Fig. 2.20).

(a) The molten pool with parent plate and weld metal mixed at temperatures above the melting point, 1500 °C.

(b) A region of BCC delta iron and austenite (FCC gamma iron) mixed.

(c) A region of austenite above the upper critical temperature, 900–1400 °C.

(d) A region of austenite and ferrite (BCC alpha iron), where ferrite is being precipitated (between upper and lower critical temperatures).

(e) The parent plate of ferrite and pearlite.

There is a high possibility, in addition, that oxygen or even nitrogen may be absorbed into the weld itself. We have seen that when oxidation occurs on the crystal boundaries, the impact strength and fatigue resistance of the metal are greatly reduced, and hence a weld which has absorbed oxygen will show these symptoms. The formation of iron nitride (Fe_4N) also makes the weld brittle. The nitride is usually present in the form of fine needle-shaped crystals visible under the high-powered microscope. The weld must be safeguarded from these defects as much as possible.

Evidently, also, the composition of the filler rod or electrode compared with that of the parent metal will be of great importance, since this will naturally alter the properties of the steel at or near the weld. If the mass of the parent metal is small and cooling is very quick, the weld may be tough and strong but brittle due to the presence of martensite and this will particularly be the case if the carbon content is high. If, however, cooling is slow, structures of varying forms of ferrite and pearlite are found, giving a lower strength and decreased hardness, but at the same time a very much increased ductility and impact strength. Evidently, therefore, the welding of a given joint requires consideration as to what properties are required in the finished weld (tenacity, ductility, impact strength, resistance to wear and abrasion, etc.). When this is settled the method of welding and the rate of cooling can be considered, together with the choice of suitable welding rods.

These considerations are of particular importance in the case of the welding of alloy steels, since great care is necessary in the choice of suitable welding rods, which will give the weld the correct properties required. In many cases, heat treatment is advisable after the welding operation, to remove internal stresses and to modify the crystal structure, and this treatment must be given to steels such as high-tensile and chrome steels. The study of welding of these steels is, however, still proceeding (Volume 2, Chapter 1).

During the welding process, the part of the weld immediately under the flame or arc is in the molten condition, the section that has just been welded

Fig. 2.20

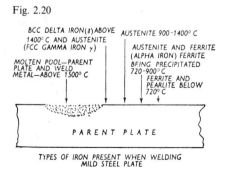

TYPES OF IRON PRESENT WHEN WELDING
MILD STEEL PLATE

is cooling down from this condition, while the section to be welded is comparatively cold. This, therefore, is virtually a small steel-casting operation, the melting and casting process taking place in a very short time and the weld metal after deposition being 'as cast' steel.

As a result we expect to find most of the various structures (martensite, bainite) that we have considered, and the point of greatest interest to the welder is, what structures will remain on cooling. Evidently the structure that remains will determine the final strength, hardness, ductility and resistance to impact of the weld. Since these structures will be greatly affected by the absorption of any elements that may be present, it will be well to consider these first.

Oxygen

Oxygen may be absorbed into the weld, forming iron oxide (Fe_3O_4) and other dioxides such as that of silicon. This iron oxide may also be absorbed into the weld from the steel of the welding rod or electrode. If iron oxide is formed it may react with the carbon in the steel to form carbon monoxide, resulting in blowholes. If this iron oxide is present in any quantity (as in the case when using bare wire electrodes in arc welding, or excess of oxygen in the oxy-acetylene process), oxidation of the weld will occur and this produces a great increase in the grain size, which is easily observed on the microphotograph. Even normalizing will not then produce a fine grain. This oxygen absorption, therefore, has a bad effect on the weld, reducing its tensile strength and ductility and decreasing its resistance to corrosion. Covered arc welding electrodes may contain deoxidizing material to prevent the formation of iron oxide, or sufficient silica to act on the iron oxide to remove it and form iron silicate (slag).

Nitrogen

The percentage of nitrogen in weld metal can vary considerably, and the results of experiments performed have led to the following conclusions:

(a) There is very low absorption by the oxy-acetylene process (maximum 0.02%).

(b) There is much greater absorption in arc welding (0.15 to 0.20%) which is influenced by (1) the current conditions that may cause the nitrogen content to vary from 0.14 to 0.2%, (2) the nature of the atmosphere. By the use of electrodes covered with hydrogen-releasing coatings, e.g. sawdust, the nitrogen content may be brought down to 0.02%.

(c) As regards the thickness of the coatings, use of a very thick covering may reduce the nitrogen from 0.15 to 0.03%.

Nitrogen is found in the weld metal trapped in blowholes (although nitrogen itself does not form the blowholes) and as crystals of iron nitride (Fe_4N), known as nitride needles. Nitrogen, however, may be in solution in the iron, and to cause the iron nitride needles to appear the weld has to be heated up to about 800–900 °C. The nitrogen tends to increase the tensile strength but *decreases* the ductility of the metal. Low-nitrogen steels are now supplied when required for deep pressing since they are less prone to cracking.

Hydrogen

Hydrogen is absorbed into mild steel weld metal during arc welding with covered electrodes. The hydrogen is present in the composition of many flux coatings and in its moisture content. It begins to diffuse out of the weld metal immediately after the welding process, and continues to do so over a long period. The presence of this hydrogen reduces the tensile strength of the weld. (An experiment to illustrate this is given in Volume 2, Chapter 1.)

Carbon

If we attempt to introduce carbon into the weld metal from the filler rod, the carbon either oxidizes into carbon monoxide during the melting operation or reacts with the weld metal and produces a porous deposit. Arc welded metal cools more quickly than oxy-acetylene welded metal and hence the former may be expected to give a less ductile weld, but the quantity of carbon introduced in arc welding (pick-up) is too small to produce brittle welds in this way.

The effect on the weld metal of the carbon contained in the parent metal is, however, important, especially when welding medium or high carbon steels. In this case, the carbon may diffuse from the parent metal, due to its relatively high carbon content, into the weld metal and form, near the line of fusion of weld and parent metal, bands of high carbon content sufficient to produce hardness and brittleness if cooled rapidly.

Structural changes

The question of change of structure during welding depends amongst other things on:

(1) The process used.
(2) The type and composition of the filler rod and, if arc welding is employed, the composition of the covering of the electrode.
(3) The conditions under which the weld is made, i.e. the amount of oxygen and nitrogen present.
(4) The composition of the parent metal.

The change of structure of the metal is also of great importance, as previously mentioned. This will depend largely upon the amount of carbon and other alloying elements present and upon the rate at which the weld cools.

In arc welding the first run of weld metal flowing onto the cold plate is virtually chill cast. The metal on the top of the weld area freezes quickly as the heat is removed from it and small chill crystals are formed. Below, these crystals grow away from the sides towards the hotter regions of the weld metal and thus growth is faster than the tendency to form new dendrites so that columnar crystals are formed.

Because of the high temperature of the molten metal in arc welding these crystals have enough time to grow. On each side of the weld is a heat-affected zone in which the temperature has been raised above the recrystallization temperature and in which, therefore, grain growth has occurred. Beyond this zone the plate structure is unaffected (Figs. 2.21, 2.24).

When a second run is placed over the first there will be:

(1) A refined area in the first run where recrystallization temperature is exceeded and the columnar crystals of the first run are reformed as small equi-axed crystals.

(2) A region between this and the parent plate where grain growth has taken place because the temperature has been well above re-crystallization temperature.

(3) The second run will form columnar crystals on its below surface layers because of the quick cooling when in contact with the atmosphere (Figs. 2.22, 2.25).

In gas welding the heat-affected zone will be wider than in arc welding because, although the flame temperature is below that of the arc, the arc is more localized and the temperature is raised more quickly. When the flame is applied to the plate its temperature is raised so that chill casting does not

Fig. 2.21

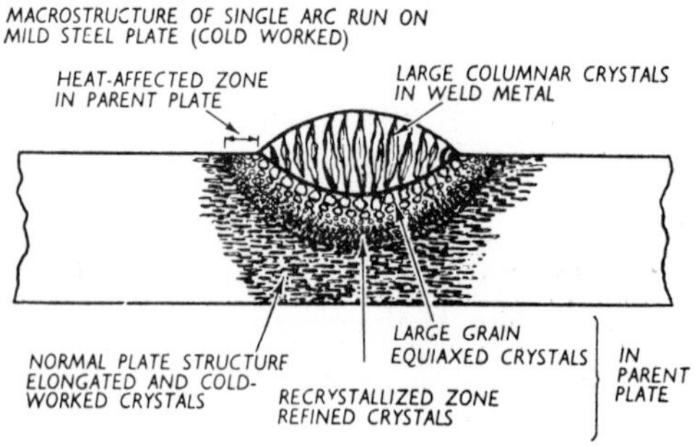

MACROSTRUCTURE OF SINGLE ARC RUN ON
MILD STEEL PLATE (COLD WORKED)

HEAT-AFFECTED ZONE
IN PARENT PLATE

LARGE COLUMNAR CRYSTALS
IN WELD METAL

NORMAL PLATE STRUCTURE
ELONGATED AND COLD-
WORKED CRYSTALS

LARGE GRAIN
EQUIAXED CRYSTALS

RECRYSTALLIZED ZONE
REFINED CRYSTALS

IN
PARENT
PLATE

occur. Grain growth, however, will be more pronounced because the heat is applied for a longer period than in arc welding (Fig. 2.23*a*, *b*).

A consideration of this subject makes very evident the reason why austenitic alloy steels present such a problem in welding. These steels owe their properties to their austenitic condition, and immediately they are subjected to the heat of the welding process, they have their structure greatly modified. It is nearly always imperative that after welding steels of this class, they should be heat treated in order to correct as far as possible this change of structure. In addition, owing to the number of alloying elements contained in these steels, it becomes very difficult to obtain a weld whose properties do not differ in a marked degree from those of the parent metal. Hence the welding of alloy steels must be considered for each particular steel and with reference to the particular requirements and

Fig. 2.22

Fig. 2.23

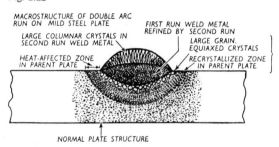

(*a*)

(*b*)

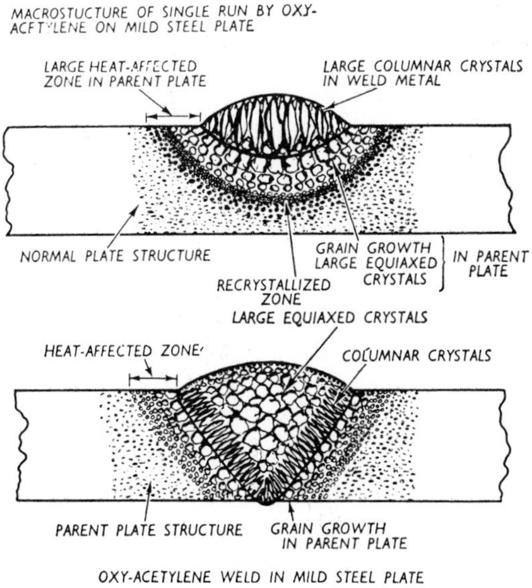

OXY-ACETYLENE WELD IN MILD STEEL PLATE

service conditions. In addition, great care must be taken in selecting a suitable electrode or filler rod.

The microscope can be used extensively to observe the effect of welding on the structure. When once the observer is trained to recognize the various structures and symptoms, the microscope provides accurate information about the state of the weld. The microscope can indicate the following points, which can hardly be found by any other method and, as previously mentioned, can also indicate fine hair cracks unperceived in X-ray

Fig. 2.24

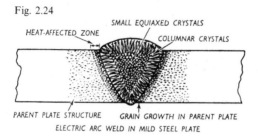

ELECTRIC ARC WELD IN MILD STEEL PLATE

Fig. 2.25. Section of a flat butt weld in rolled steel plate (× 3). Macrographs of the regions labelled 1, 2, 3 and 4 are shown in Figs. 2.26 and 2.27.

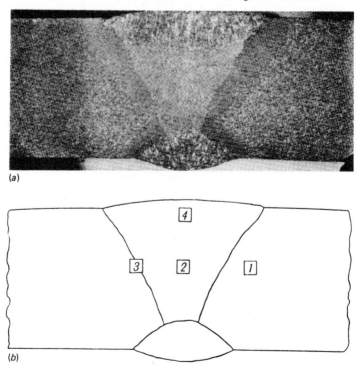

photographs, together with any slag inclusions and blowholes of microscopic proportions. Faults in structure indicated by the microscope using various magnifications are:

(1) True depth of penetration of the weld as indicated by the crystal structure.

(2) The actual extent of the fusion of weld and parent metal (Fig. 2.24).

(3) The actual structure of the weld metal and its conditions (Figs. 2.24, 2.25, 2.26, 2.27).

(4) The area over which the distribution due to the heating effect of the welding operation has occurred.

(5) The amount of nitrogen and oxygen absorption, as seen by the presence of iron oxide and iron nitride crystals.

Dilution

When two metals are fusion welded together by metal arc, TIG, MIG or submerged arc processes, the final composition consists of an admixture of parent plate and welding wire. The parent plate has melted in with the filler and has diluted it and this dilution may be expressed as a percentage thus:

$$\text{percentage dilution} = \frac{\text{weight of parent metal in weld}}{\text{total weight of weld.}} \times 100.$$

Fig. 2.26
(1) Rolled steel plate. × 75.
(2) Weld metal showing refined structure. × 75.

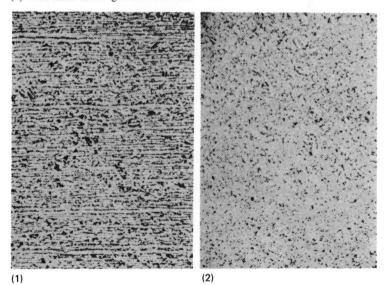

(1)   (2)

If there are 15 parts by weight of parent plate in 75 parts by weight of weldmetal then the dilution is $15/75 \times 100 = 20\%$.

Average values of dilution for various processes are:

Metal arc	25–40%
Submerged arc	25–40%
MIG (spray transfer)	25–50%
MIG (dip transfer)	15–30%
TIG	25–50%

and it can be seen that minimum dilution is obtained using MIG (dip

Fig. 2.27
(3) Junction of weld metal and plate. × 75.
(4) Top layer. × 75.

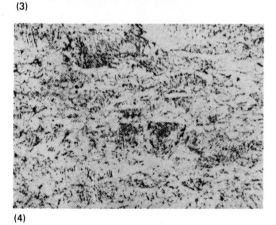

(3)

(4)

transfer). Many factors affect dilution. Evidently with a single-run weld there will be greater dilution than with multi-runs and there is always considerable dilution in any root run. The greater the amount of weaving the greater the dilution.

When dissimilar metals are to be welded together the final weld will suffer dilution from each parent plate and for a successful weld it must do so without major defects, including liability of cracking. In addition the physical and mechanical properties and resistance to corrosion must be as near as possible to those of the parent plate. As all these criteria are not generally obtainable a welding wire must be chosen which gives the optimum properties for a given situation.

Example

A plate of 9% Ni steel is welded with an inconel welding wire of composition 80% Ni–20% Cr. What will be the approximate composition of the final weld if there is 40% dilution?

With 40% dilution the plate will contribute 40% and the welding wire 60%.

	Nickel	Iron	Chromium
9% nickel plate	$\dfrac{40}{100} \times 9 = 3.6$	$\dfrac{40}{100} \times 91 = 36.4$	
80% Ni–20% Cr welding wire	$\dfrac{60}{100} \times 80 = 48$		$\dfrac{60}{100} \times 20 = 12$
	51.6	36.4	12

Therefore the approximate composition of weld metal is:

$$51.5\% \text{ Ni, } 36.5\% \text{ Fe, } 12\% \text{ Cr.}$$

A plate of monel (70% Ni–30% Cu) is to be welded to a plate of stainless steel (18% Cr–12% Ni–70% Fe) using an Incoweld A wire (75% Ni–15% Cr–8% Fe). Assuming 30% dilution, what will be the approximate composition of the final weld?

With 30% dilution each plate will contribute 15% and the welding wire 70%.

	Nickel	Chromium	Copper	Iron
Monel plate	$\dfrac{15}{100} \times 70 = 10.5$		$\dfrac{15}{100} \times 30 = 4.5$	
Stainless steel plate	$\dfrac{15}{100} \times 12 = 1.8$	$\dfrac{15}{100} \times 18 = 2.7$		$\dfrac{15}{100} \times 70 = 10.5$
Welding wire	$\dfrac{70}{100} \times 75 = \dfrac{52.5}{64.8}$	$\dfrac{70}{100} \times 15 = \dfrac{10.5}{13.2}$	4.5	$\dfrac{70}{100} \times 8 = \dfrac{5.6}{16.1}$

Therefore the approximate composition of weld metal is:

$$65\% \text{ Ni, } 13\% \text{ Cr, } 4.5\% \text{ Cu, } 16\% \text{ Fe.}$$

Pick-up. This is the term applied to the absorption or transfer of elements from parent plate or non-consumable electrode into the weld metal and is closely associated with dilution. When overlaying plain low-carbon steels with nickel-base alloys there is a tendency for the weld metal to pick up iron from the parent plate, resulting in a lowering of corrosion resistance of the overlay. The pick-up must be kept as low as possible. When welding cast iron with the metal arc processes carbon pick-up can lead to the undesirable excessive precipitation of carbides, which are hard and brittle. Tungsten pick-up can occur when welding with the gas-shielded tungsten electrode (TIG) process when excessive currents are used, resulting in the pick-up of tungsten particles from the electrode into the weld metal (vol. 2, ch. 3).

Cracking in steel

A crack is a fissure produced in a metal by tearing action. Hot or solidification cracking is caused in the weld metal itself by tearing of the grain boundaries before complete solidification has taken place and while the metal is still in the plastic state. The crack may be continuous or discontinuous and often extends from the weld root and may not extend to the face of the weld. Cold cracking occurs in both weld metal and adjacent parent plate (HAZ) and may be due to excessive restraint on the joint, insufficient cross-sectional area of the weld, presence of hydrogen in the weld metal or embrittlement in the HAZ of low alloy steels.

Factors which may promote hot cracking

Current density: a high density tends to promote cracking.
The distribution of heat and hence stress in the weld itself.
Joint restraint and high thermal severity.
Crack sensitivity of the electrode.
Dilution of the weld metal.
Impurities such as sulphur, and high carbon or nickel content.
Pre-heating, which increases the liability to cracking.
Weld procedure. Higwelding speeds and long arc increase
    sensitivity and crater cracking indicates a crack sensitivity.

Factors which may promote cold cracking

Joint restraint and high thermal severity.
Weld of insufficient sectional area.
Hydrogen in the weld metal.
Presence of impurities.
Embrittlement of the HAZ (low-alloy steels).
High welding speeds and low current density.

The effect of deformation on the properties of metals

Cold working

When ductile metal is subjected to a stress which exceeds the elastic limit, it deforms plastically by an internal shearing process known as *slip*.

Plastic deformation is permanent so that when the applied stress is removed the metal remains deformed and does not return to its original size and shape as is the case with *elastic* deformation.

Plastic deformation occurs in all shaping processes such as rolling, drawing and pressing, and may occur locally in welded metals owing to the stresses set up during heating and cooling. When plastic deformation is produced by cold working it has several important effects on the structure and properties of metal:

(1) The metal grains are progressively elongated in the direction of deformation.

(2) With heavy reduction by cold work the structure becomes very distorted, broken up, and fibrous in character.

(3) If there is a second constituent present, as in many alloys, this becomes drawn out in threads or strings of particles in the direction of working, thus increasing the fibrous character of the material.

(4) The deformation of the structure is accompanied by progressive hardening, strengthening and loss of ductility and by an increased resistance to deformation. Cold worked metals and alloys are therefore harder, stronger and less ductile than the same materials in the undeformed state. The properties of cold worked materials may also differ in different directions owing to the fibrous structure produced.

(5) If the deformation process does not act uniformly on all parts of the metal being rolled or drawn, then internal stresses may be set up. These stresses may add to a subsequent working stress to which the metal is subjected, and they also render many metals and alloys subject to a severe form of intercrystalline corrosion, known as stress-corrosion; for example, the well-known season cracking of cold worked brass.

(6) If the deformation process is carried to its limit, the metal loses all of its ductility and breaks in a brittle manner. Brittle fracture of a similar kind can be produced in metals without preliminary deformation if they are subjected to certain complex stress systems which prevent deformation by slipping (three tensile stresses acting perpendicularly to each other). Such conditions can occur

in practice in the vicinity of a notch and are sometimes set up in the regions affected by welding.

The effect of heat on cold worked metals : annealing and recrystallization

On heating a cold worked metal, the first important effect produced is the relief of internal stress. This occurs without any visible change in the distorted structure. The temperature at which stress relief occurs varies from 100 °C up to 500 °C according to the particular metal concerned. In general, the higher the melting point, the higher is the temperature for this effect.

On further heating the cold worked metal, no other changes occur until a critical temperature known as the 'recrystallization temperature' is reached. At the recrystallization temperature, the distorted metal structure is able to re-arrange itself into the normal unstrained arrangement by recrystallizing to form small equi-axed grains. The mechanical properties return again to values similar to those which the metal possessed before the cold working operation, i.e. hardness and strength fall, and ductility increases. However, if the extent of the distortion was very severe (suppose for example there had been a 60% or more reduction by rolling) the recrystallized metal may exhibit different properties in different directions.

If the metal is heated to a higher temperature than the minimum required for recrystallization to occur, the new grains grow progressively larger and the strength and working properties deteriorate.

The resoftening which accompanies recrystallization is made use of during commercial cold working processes to prevent the metal becoming too brittle, as, for example, after a certain percentage of reduction by cold rolling, the metal is annealed to soften it, after which further reduction by cold working may be done.

The control of both working and annealing operations is important because it affects the grain size, which, in turn, controls the properties of the softened material. It is generally advantageous to secure a fine grain size. The main factors determining the grain size are:

(1) The prior amount of cold work – the recrystallized grain size decreases as the amount of prior cold work increases.

(2) The temperature and time of the annealing process. The lowest annealing temperature which will effect recrystallization in the required time produces the finest grain size.

(3) Composition. Certain alloying elements and impurities restrict grain growth.

The temperature at which a cold worked metal or alloy will recrystallize depends on:

(1) Its melting point. The higher the melting point the higher the recrystallization temperature.

(2) Its composition and constitution. Impurities or alloying elements present in solid solution raise the recrystallization temperature. Those present as second constituents have little effect (although these are the type which tend to restrict grain growth once the metal or alloy has recrystallized).

(3) The amount of cold work. As the extent of prior cold work increases, so the recrystallization temperature is lowered.

(4) The annealing time. The shorter the time of annealing the higher the temperature at which recrystallization will occur.

These factors must all be taken into account in determining practical annealing temperatures.

Hot working

Metal and alloys which are not very ductile at normal temperatures, and those which harden very rapidly when cold worked, are generally fabricated by hot working processes, namely forging, rolling, extrusion, etc. Hot working is the general term applied to deformation at temperatures above the recrystallization temperature. Under these conditions the hardening which normally accompanies the deformation is continually offset by recrystallization and softening. Thus a hot worked material retains an unstrained equi-axed grain structure, but the size of the grains and the properties of the structure depend largely on the temperature at which working is discontinued. If this is just above the recrystallization temperature, the grains will be unworked, fine and uniform, and the properties will be equivalent to those obtained by cold working followed by annealing at the recrystallization temperature.

If working is discontinued well above the recrystallization temperature, the grains will grow and develop inferior properties, while a duplex alloy may develop coarse plate structures similar to those present in the cast state and have low shock-strength.

On the other hand, if working is continued until the metal has cooled below its recrystallization temperature, the grains will be fine but will be distorted by cold working, and the metal will therefore be somewhat harder and stronger but less ductile than when the working is discontinued at or above the recrystallization temperature.

Insoluble constituents in an alloy become elongated in the direction of work as in cold working and produce similar directional properties.

Hot working is used extensively for the initial 'breaking down' of large ingots or slabs, even of metals which can be cold worked. This is because of the lower power required for a given degree of reduction by hot work. Hot work also welds up clean internal cavities but it tends to give inferior surface properties since some oxide scale is often rolled into the surface. Further, it does not permit such close control of finishing gauges, and cannot be used effectively for finishing sheet or wire products.

Impurities which form low melting point constituents can ruin the hot working properties of metals and alloys, for example excess sulphur in steel.

Cold working as a major fabrication process is restricted to very ductile materials such as pure and commercial grades of copper, aluminium, tin and lead, and to solid solution alloys, such as manganese–aluminium alloys; brasses, containing up to 35% zinc; bronzes containing up to 8% tin; copper–nickel alloys; aluminium bronzes, containing up to 8% aluminium; nickel silvers (copper–nickel–zinc alloys) and tin–base alloys such as pewter. It is also used in the finishing working stages of many other metals and alloys to give dimensional control, good surface quality and the required degree of work hardening.

Iron, steels, high zinc or high aluminium copper alloys, aluminium alloys such as those with copper, magnesium and zinc, pure zinc and pure magnesium, and the alloys of these metals are all generally fabricated by hot working, though cold working may be used in the late stages, especially for the production of wire or sheet products.

Iron and steel

Pure iron is very ductile and can be both hot and cold worked.

The carbon content of steel makes it less ductile than iron. Dead mild steel of up to 0.15% carbon is very suitable for cold working and can be flanged and used for solid drawn tubes. Although it is slightly hardened by cold work the modulus of elasticity is unaffected. If steel is hot worked at a temperature well above the recrystallization temperature, grain growth takes place and the impact strength and ductility are reduced, hence it should be worked at a temperature just above the recrystallization point, which is 900–1200°C for mild steel and 750–900°C for high-carbon steel. Thus, in general, hot working increases the tensile and impact strength compared with steel in the cast condition, and as the carbon content increases, the steel become less ductile and it must be manipulated by hot working.

Copper

Copper is very suitable for cold working as its crystals are ductile and can suffer considerable distortion without fracture, becoming harder, however, as the amount of cold work increases. Since welding is performed above the annealing temperature it removes the effect of cold work.

Copper is very suitable for hot working and can be hot rolled, extruded and forged.

Brasses

Brasses of the 85/15 and 70/30 composition are very ductile. Brass can be heavily cold worked without suffering fracture. This cold work modifies the grain structure, increases its strength and hardens and gives it varying degrees of temper.

Since these brasses are so easily cold worked there is little advantage to be gained by hot work which however, is quite suitable.

For brass of the 60/40 type ($\alpha + \beta$ structures) see pp. 120–1.

The β structure, which is zinc rich, is harder than the α structure and will stand little cold work without fracture. If the temperature is raised, however, to about 600 °C it becomes more easily worked.

Brasses of this composition therefore should be hot worked above 600–700°C giving a fine grain and fibrous structure. Below 600°C for this structure can be considered cold work.

Copper–nickel alloys

80/20 copper-nickel alloy is suitable for hot and cold work but is particularly suitable for the latter due to its extreme ductility.

Nickel–chrome alloys are generally hot worked.

Bronzes. Gunmetal (Cu 88%, Sn 10%, Zn 2%) must be hot worked above 600 °C and not cold worked. Phosphor bronze with up to 6–7% tin can be cold worked.

Aluminium bronze is similar to brass in being of two types:

(1) α structure containing 5–7% Al, 93% Cu.

(2) $\alpha + \beta$ structure containing 10% Al, 90% Cu.

(1) The α structure, which is the solid solution of aluminium in copper, is ductile and this type is easily worked hot or cold.

(2) The $\alpha + \beta$ structure is rendered more brittle by the presence of the β constituent, which becomes very hard and brittle at 600 °C due to the formation of another constituent which, in small quantities, increases the tensile strength; hence this alloy must be hot worked.

Aluminium. Pure aluminium can be worked hot or cold but weldable alloys of the work hardening type such as Al–Si, Al–Mn and Al–Mn–Mg, are hardened by cold work which gives them the required degree of temper and they soften at 350 °C. Duralumin is hot short above 470 °C and too brittle to work below 300 °C so it should be worked in the range 400–470 °C. Y alloy can be hot worked.

Non-ferrous metals

Copper

Copper is found in the ore copper pyrites ($CuFeS_2$) and is first smelted in a blast or reverberatory furnace, and is then in the 'blister' or 'Bessemer' stage. In this form it is unsuitable for commercial use, as it contains impurities such as sulphur and oxygen. Further refining may be carried on by the furnace method, in which oxidation of the sulphur and other impurities occurs, or by an electrical method (called electrolytic deposition), resulting in a great reduction of the impurities.

In the refining and melting processes, oxidation of the copper occurs and the excess oxygen is removed by reducing conditions in the furnace. This is done by thrusting green wooden poles or tree trunks under the surface of the molten copper, which is covered with charcoal or coke to exclude the oxygen of the air. The 'poling', as it is called, is continued until the oxygen content of the metal is reduced to the limits suitable for the work for which the copper is required.

The oxygen content of the copper is known as the 'pitch' and poling is ended when the 'tough pitch' condition is reached.

Oxygen in copper. In this condition the oxygen content varies from 0.025 to 0.08%. The oxygen exists in the cast copper as minute particles of cuprous oxide (Cu_2O) (Fig. 2.28a).

The amount of oxygen in the copper is most important from the point of view of welding, since the welding of copper is rendered extremely difficult by the presence of this copper oxide. When molten, copper oxide forms a eutectic with the copper and this collects along the grain boundaries, reducing the ductility and increasing the tendency to crack. Any hydrogen present, as occurs when there are reducing conditions in the flame, reduces the copper oxide to copper and water is also formed. This is present as steam, which causes porosity and increases the liability of cracking. For welding purposes, therefore, it is much preferable to use copper almost free from oxygen, and to make this, deoxidizers such as phosphorus, silicon, lithium, magnesium, etc., are added to the molten metal, and they combine with oxygen to form slag and thus *deoxidize* the copper (Fig. 2.28*b*). The welding of 'tough pitch' copper depends so much on the skill of the welder that it is always advisable to use 'deoxidized copper' for welding, and thus eliminate any uncertainty.

Arsenic in copper. When arsenic up to 0.5% is added to copper, the strength and toughness is increased. In addition to this, it increases its resistance to fatigue and raises by about 100 °C the temperature at which softening first occurs and enables it to maintain its strength at higher temperatures. Arsenic is undesirable in copper intended specifically for welding purposes, since it makes welding more difficult. Arsenical copper can be welded by the same method as for ordinary copper, and if care is taken the welds are quite satisfactory. As with ordinary copper, it may be obtained in the deoxidized or the tough pitch form, the former being the more suitable for welding.

Properties of copper. Copper is a red-coloured metal having a melting point of 1083 °C and a density of 8900 kg/m³. The mechanical properties of copper depend greatly on its condition, that is, whether it is in the 'as cast' condition or whether it has been hot or cold worked, hammered, rolled, pressed, or forged.

The tensile strength 'as cast' is about 160 N/mm². Hot rolling and forging, followed by annealing, modifies its structure and increases its strength to about 220 N/mm². Cold working by hammering, rolling, drawing and pressing hardens copper and raises its tensile strength, but it becomes less ductile.

Fig. 2.28
(*a*) Cuprous oxide in copper. × 100.
(*b*) Deoxidized copper. × 75.

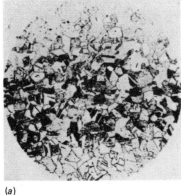

(a)   (b)

Very heavy cold worked copper may have a tensile strength equal to that of mild steel, but it has very little ductility in this state.

The temper of copper. Copper is tempered by first getting it into a soft or annealed condition, and then the temper required is obtained by cold working it (hammering, rolling, etc.). Thus it is the reverse process from the tempering of steel. Soft-temper copper is that in the annealed condition. It has a Brinell hardness of about 50. After a small amount of cold working, it becomes 'half hard' temper, and further cold working makes it hard temper having a Brinell figure of about 100. Intermediate hardness can of course be obtained by varying the amount of cold working.

Annealing. Copper becomes hard and its structure is deformed when cold worked, and annealing is therefore necessary to soften it again. To anneal the metal, it is usual to heat it up to about 500 °C, that is, dull red heat, and either quench it in water or let it cool out slowly, since the rate of cooling does not affect the properties of the pure metal. Quenching, however, removes dirt and scale and cleans the surface. The surface of the copper can be further cleaned or 'pickled' by immersing it in a bath of dilute sulphuric acid containing 1 part of acid to 70 parts of water. If nitric acid is added, it accelerates the cleaning process. If copper has a surface polish, heating to the annealing temperature will cause the surface to scale, which is undesirable; hence annealing is usually carried out in a non-oxidizing atmosphere in this case.

Crystal or grain size. Under the microscope, cold worked copper shows that the grains or crystals of the metal have suffered distortion. During the annealing process, as with steel, recrystallization occurs and new crystals are formed. As before, if the annealing temperature is raised too high or the annealing prolonged too long, the grains tend to grow. With copper, however, unlike steel, the rate of growth is slow, and this makes the annealing operation of copper subject to a great deal of latitude in time and temperature. This explains why it is immaterial whether the metal cools out quickly or slowly after annealing.

The main grades in which copper is available are: (1) oxygen-bearing (tough pitch) high conductivity; (2) oxygen-free high conductivity; (3) phosphorus deoxidized.

Alloys of copper

The alloys of copper most frequently encountered in welding are: copper–zinc (brasses and nickel silvers); copper–tin (phosphor bronzes and gunmetal); copper–silicon (silicon bronzes); copper–aluminium (aluminium bronzes); copper–nickel (cupro-nickels); and heat-treatable alloys such as copper–chromium and copper–beryllium. These are also discussed in the section on the welding of copper by the TIG process.

Filler metals for gas-shielded arc welding of copper (conforming to BS 2901)

BS2901 designation	Nominal composition (%)
C7	0.15–0.35 Mn; 0.20–0.35 Si; Rem. Cu
C8	0.1–0.3 Al; 0.1–0.3 Ti; Rem. Cu
C9	0.75–1.25 Mn; 2.75–3.25 Si; Rem. Cu
C10	0.02–0.40 P; 4.5 Sn; Rem. Cu
C11	6.0–7.5 Sn; 0.02–0.40 P; Rem. Cu
C12	6.0–7.5 Al; 1.0–2.5 (Fe + Mn + Ni); Rem. Cu
C12 Fe	6.5–8.5 Al; 2.5–3.5 Fe; Rem. Cu
C13	9.0–11.0 Al; 0.75–1.5 Fe; Rem. Cu
C16	10.0–12.0 Ni; 0.20–0.50 Ti; 1.5–1.8 Fe; 0.5–1.0 Mn; Rem. Cu
C18	30.0–32.0 Ni; 0.20–0.50 Ti; 0.40–1.0 Fe; 0.5–1.5 Mn; Rem. Cu
C20	8.0–9.5 Al; 1.5–3.5 Fe; 3.5–5.0 Ni; 0.5–2.0 Mn; Rem. Cu
C21	0.02–0.10 B; Rem. Cu
C22	7.0–8.5 Al; 2.0–4.0 Fe; 1.5–3.5 Ni; 11.0–14.0 Mn; Rem. Cu

Brasses or copper–zinc alloys. Zinc will dissolve in molten copper in all proportions and give a solution of a uniform character. Uniform solution can be obtained when solidified if the copper content is not less than about 60%. For example, 70% copper–30% zinc consists of a uniform crystal structure known as 'alpha' (α) solid solution and is shown in Fig. 2.29*a*. If the percentage of zinc is now increased to about 40%, a second constituent structure, rich in zinc, appears, known as 'beta' (β) solid solution, and these crystals appear as reddish in colour, and the brass now has a duplex structure as shown in Fig. 2.29*b*. These crystals are hard and increase the tensile strength of the brass but lower the ductility. The alpha-type brass, which can be obtained when the copper content has a minimum value of 63%, has good strength and ductility when cold and is used for sheet, strip, wire and tubes. The alpha–beta, e.g. 60% copper–40% zinc, is used for casting purposes, while from 57 to 61% copper types are suitable for hot rolling, extruding and stamping. Hence a great number of alloys of varying copper–zinc content are available. Two groups, however, are of very great importance, as they occur so frequently:

(1) Cartridge brass: 70% copper and 30% zinc, written 70/30 brass.
(2) Yellow or Muntz metal: 60% copper and 40% zinc, written 60/40 brass.

The table illustrates the composition and uses of various copper–zinc and copper–tin alloys.

Properties of brass. Brass is a copper-zinc alloy with a golden colour which can be easily cast, forged, rolled, pressed, drawn and machined. It has a good resistance to atmospheric and sea-water corrosion and therefore is used in the manufacture of parts exposed to these conditions. As the copper content in the brass is decreased, the colour changes from the reddish colour of copper to yellow and then pale yellow.

The density varies from 8200 to 8600 kg/m³, depending on its composition. The heat and electrical conductivity decrease greatly as the zinc content increases, and the melting point is lowered as the copper content decreases, being about 920 °C for 70/30 brass.

Brass for brazing purposes can vary greatly in composition, depending upon the melting point required; for example, three brazing rod compositions are: 54% copper, 46% zinc; 50% copper, 50% zinc (melting at

Fig. 2.29. (*a*) Rolled and annealed cartridge brass (70/30). This brass has a simple structure, containing only crystals of alpha solid solution, that is zinc dissolved in copper. × 100. (*b*) Rolled and annealed yellow metal (60/40). This brass is a mixture of alpha crystals (white areas), and beta crystals (black areas), richer in zinc. × 100.

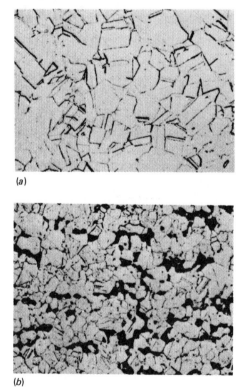

(a)

(b)

860 °C); and 85% copper, 15% zinc. The choice of the alloy therefore depends on the work for which it is required. For welding brass, the filler rod usually contains phosphorus or silicon, which act as deoxidizing agents, that is, they remove any oxygen from the weld.

As the copper content is reduced, there is a slight increase in the tensile strength.

Annealing. Examination of cold worked brass under the microscope shows that, as with copper, distortion of the crystals has taken place. When its temperature is raised to about 600 °C, recrystallization takes place and the crystals are very small. The rate of growth depends on the temperature, and the higher the temperature the larger the crystals or grains. The annealing time (as with steel) also affects their size. In over-annealed brass, having

Composition of copper–zinc and copper–tin alloys

% Composition by weight			Other elements	Uses
Cu	Zn	Sn		
90	10			Gilding metal.
88	2	10		Admiralty gun metal. Good casting properties and corrosion resistance. Widely used for pumps, valves, etc.
85	15			Architectural and decorative work.
80	20			
70	30			Cartridge brass. For deep drawing and where high strength and ductility are required.
66	34			2/1 brass.
62–65	38–35			Common brass.
60	40			Yellow or Muntz metal. Works well when hot. For brass sheets and articles not requiring much cold work during manufacture.
76	22		2 Al	Aluminium brass.
61–63.5	rem	1–1.4		Naval brass.
70–73	rem	1–1.5	0.02–0.06 As	Admiralty brass.
63–66	rem		0.75–1.5 Pb ⎱	Leaded brass, casts well. Easily hot stamped and extruded. Machines well.
61–64	rem		1.0–2.0 Pb ⎬	
58–60	rem		1.5–2.5 Pb ⎰	
58	38		Mn, Fe, Ni or Sn approx 4%	Manganese bronze, high-tensile alloys for castings and bearings.
Rem			2.7–3.5 Si 0.75–1.25 Mn	Silicon bronze.
Rem		3–4.5	0.02–0.4 P	3% phosphor bronze.
Rem		4.5–6.0	0.02–0.4 P	5% phosphor bronze.
Rem		6.0–7.5	0.02–0.4 P	7% phosphor bronze.

Note. Material can be supplied as annealed (O); various tempers produced by cold work and partial annealing and indicated by $\frac{1}{4}$H, $\frac{1}{2}$H, etc.; spring tempers (SH and ESH); solution treated (W) and precipitation hardened (P).

large crystals, they may show up on the surface after cold working as an 'orange-peel' effect.

Annealing at too high a temperature may also cause pitting or deterioration of the surface by scaling.

Brass can be either quenched out in water or allowed to cool out slowly after annealing. If quenched, the surface scale is removed, but care must be taken with some brasses lest the ductility suffers.

Temper. Brass is tempered in the same way as copper, that is, by cold working. In the annealed condition it is 'soft temper' (60–80 Brinell). A little cold working brings it to 'half-hard temper' and further work gives it a 'hard temper' (150–170 Brinell). More cold working still, produces a 'spring-hard temper' with a Brinell number of 170–180.

Elasticity. The tensile strength of brass varies with the amount of cold working, and it is sufficiently elastic to allow of its being used as springs when in the spring-hard temper condition.

Bronzes or copper–tin alloys

Gunmetals are zinc-containing bronzes, e.g. 88% Cu, 10% Sn, 2% Zn. Wrought phosphor bronzes contain up to 8% tin and up to 0.4% phosphorus, while cast forms contain at least 10% tin with additions of lead to promote free machining and pressure tightness.

Gunmetal was chiefly used, as its name suggests, for Admiralty and Army Ordnance work, but is now used chiefly where resistance against corrosion together with strength is required. Lead bronze has lead added to improve its properties as a bearing surface and to increase its machinability.

Phosphor bronze has largely replaced the older bearing bronze for bearings owing to its increased resistance to wear. Phosphorus, when added to the copper–tin bronze, helps greatly to remove the impurities, since it is a powerful reducing agent and the molten metal is made much purer.

Bronze welding rods of copper–tin and copper–zinc composition are very much used as filler rods and electrodes in welding. They can be used for the welding of steel, cast iron, brass, bronze and copper, and have several advantages in many cases over autogenous welding since, because of their lower melting point, they introduce less heat during the welding operation. Manganese bronze can be considered as a high-tensile brass.

Nickel and nickel alloys

Nickel is a greyish-white metal melting at 1450 °C, has a specific gravity of 8.8, and has a coefficient of linear expansion of 0.000 013 per degree C.

Nickel and nickel alloys

Material	Major constituents	Applications
Nickel	almost pure nickel	High resistance to corrosion in contact with caustic alkalis, dry halogen gases and organic compounds generally.
Monel	nickel, copper	Has good corrosion resistance with good mechanical properties. A variation responds to thermal hardening of the precipitation type and has good corrosion resistance, with the mechanical properties of heat-treatable alloy steels.
Inconel	nickel, chromium, iron.	Is oxidation-resistant at high temperatures with good mechanical properties and is resistant to food acids. Widely used for heat treatment and furnace equipment.
	nickel, chromium, iron with additions of molybdenum and niobium.	High level of mechanical properties without the need for heat treatment. Good oxidation resistance and resists corrosive attack by many media. Other variations are age-hardenable with high strength at elevated temperatures, have outstanding weldability and can be welded in the heat-treated condition.
Incoloy	nickel, chromium, iron.	Oxidation-resistant at elevated temperatures with good mechanical properties. Variations with a lower silicon content used for pyrolysis of hydrocarbons as in cracking or reforming operations in the petroleum industry.
	nickel, chromium, iron with copper and molybdenum additions	Resistant to hot acid and oxidizing conditions, e.g. nitric–sulphuric–phosphoric acid mixtures.
Nimonic	nickel, chromium. nickel, chromium, cobalt. nickel, chromium, iron.	Used for gas turbine parts, heat treatment and other purposes where both oxidation resistance and high-temperature mechanical properties are required.
Brightray	nickel, chromium. nickel, chromium, iron.	Heating elements of electric furnaces, etc.
Nilo	nickel, iron.	These have a controlled low and intermediate coefficient of thermal expansion and are used in machine parts, thermostats and glass-to-metal seals.

It resists caustic alkalis, ammonia, salt solutions and organic acids, and is used widely in chemical engineering for vats, stills, autoclaves, pumps, etc. When molten, it absorbs (1) carbon, forming nickel carbide (Nl_3C), which forms graphite on cooling; (2) oxygen, forming nickel oxide (NiO), which makes the nickel very brittle, and (3) sulphur, forming nickel sulphide (NiS).

Magnesium and manganese are added to nickel in order to deoxidize it and render it more malleable.

Nickel is widely used as an alloying element in the production of alloy steels, improving the tensile strength and toughness of the steel, and is used in cast iron for the same purpose. In conjunction with chromium it gives the range of stainless steels.

Copper–nickel alloys

Nickel and copper are soluble in each other in all proportions to give a range of cupro-nickels which are ductile and can be hot and cold worked. The more important alloys are the following:

90% Cu, 10% Ni, used for heat exchangers for marine, power, chemical and petrochemical use; feed water heaters, condensers, evaporators, coolers, radiators, etc.

80% Cu, 20% Ni, used for heat exchangers, electrical components, deep-drawn pressings and decorative parts.

70% Cu, 30% Ni, which has the best corrosion resistance to sea and other corrosive waters and is used for heat exchangers and other applications given for 90/10% alloy.

The alloys used for resistance to corrosion usually have additions of iron (0.5–2.0%) and manganese (0.5–1.5%), while alloys for electrical uses are free from iron and have only about 0.2% Mn. The actual composition of Monel is 29% Cu, 68% Ni, 1.25% Fe, 1.25% Mn.

Other alloys are: 75% Cu, 25% Ni, used for coinage; and nickel silvers, which contain 10–30% Ni, 55–63% Cu with the balance zinc and are extensively used for cutlery and tableware of all kinds, being easily electro-plated (EPNS). Nickel is added to brass and aluminium bronze to improve corrosion resistance.

Aluminium

Aluminium is prepared by electrolysis from the mineral *bauxite*, which is a mixture of the oxides of aluminium, silicon and iron.

The aluminium oxide, or alumina as it is called, is made to combine with caustic soda to form sodium aluminate, thus:

$$\text{alumina} + \text{caustic soda} \rightarrow \text{sodium aluminate} + \text{water}$$
$$Al_2O_3 + 2NaOH \rightarrow 2NaAlO_2 + H_2O$$

This solution of sodium aluminate is diluted and filtered to remove iron oxide, and aluminium hydroxide is precipitated. This is dried and calcined leaving aluminium oxide (Al_2O_3). The aluminium oxide is placed together with cryolite (Na_3AlF_6) and sometimes fluorspar (CaF_2) into a cell lined with carbon, forming the cathode or negative pole of the direct current circuit. Carbon anodes form the positive pole and hang down into the mixture. The p.d. across the cell is about 6 volts and the current which passes through the mixture and fuses it may be up to 100 000 amperes. As the current flows, the cryolite is electrolysed, aluminium is set free and is tapped off from the bottom of the cell, while the fluorine produced reacts with alumina, forming aluminium fluoride again. Because such large currents are involved, aluminium production is carried out near cheap sources of electrical power.

Aluminium prepared in this way is between 99 and 99.9% pure, iron and silicon being the chief impurities. In this state it is used for making sheets for car bodies, cooking utensils, etc., and for alloying with other metals. In the 99.5%-and-over state of purity it is used for electrical conductors and other work of specialized nature.

Properties of aluminium. Pure aluminium is a whitish-coloured metal with a density of 2.6898 g/cm³ (2700 kg/m³), that is, it weighs less, volume for volume, than one-third the weight of copper (8.96 g/cm³) and just more than one-third that of iron (7.9 g/cm³). Its melting point is 660 °C (boiling point 2480 °C) and its tensile strength varies from 60 to 140 N/mm² (MPa) according to the purity and the amount of cold work performed on it.

It casts well, has high ductility and can be hammered and rolled into rod and sheet form, etc., and extruded into wire. In contact with air a thin film of oxide (alumina) forms on the surface. In non-corrosive conditions the film is thin and almost invisible but in corrosive conditions it may appear as a grey-coloured coat helping to prevent further corrosion. This oxide is removed by fluxes in oxy-acetylene welding but nowadays welding is easily performed by the TIG or MIG processes using inert gas shielding (argon or helium), by electron beam and laser, and most alloys can be resistance welded. Some alloys can be brazed using an aluminium–silicon filler rod containing up to 10% silicon.

Aluminium is a good conductor of heat and electricity and pure aluminium may be drawn into wire for power cables. For transmission lines the strands are wound over a stranded steel core to give the required strength. Weight for weight aluminium is a better conductor than copper.

The process known as anodising produces a relatively hard and thick film of oxide on the surface. The film can be dyed any colour and the thicker the film produced, the greater the corrosion resistance.

Aluminium and its alloys. Aluminium and its alloys can be divided into two groups:

(1) casting alloys
(2) wrought alloys

It is these latter which are of greatest importance to the welder. Wrought alloys may be further divided into two groups according to the treatment required in order to improve their mechanical properties: (*a*) heat-treatable; (*b*) non-heat-treatable or work-hardening; and cold work including rolling, drawing and stamping, etc., improves the mechanical properties of these alloys and increases the tensile strength and hardness. The aluminium–silicon alloys containing 10–15% silicon (a eutectic is formed with 11.7% silicon) are very fluid when molten, cast well and have considerable strength, ductility and resistance to corrosion, with small contraction on solidification. The addition of magnesium, manganese or both gives a range of work hardening alloys, while the addition of copper, magnesium and silicon gives a range of heat-treatable alloys which age-harden due to the formation of intermetallic compounds, and they have a high strength–weight ratio. The aluminium–zinc–magnesium–copper alloys, after full heat treatment give the highest strength–weight ratio.

Heat treatment. The purpose of heat treatment is to increase strength and hardness of the alloy. By the addition of small percentages of elements such as copper, magnesium, silicon, zinc, etc., intermetallic compounds are formed and the mechanical properties of the alloy are improved. Some alloys are used in the 'as cast' or wrought condition while others are modified by heat treatment and/or cold working.

Solution treatment. The heat-treatable alloy is raised in temperature to between 425 °C and 540 °C depending upon the alloy. The relatively hard constituents formed by the addition of the alloying elements are taken into solid solution and the alloy is then quenched (e.g. in water). The constituents remain in solid solution but the alloy is soft and unstable. As time passes the constituents precipitate into a more uniform pattern in the alloy and the strength and hardness increase. This is known as natural age-hardening or precipitation hardening and can be from a few hours to many months. Any forming should be performed immediately after quenching, before any age-hardening has had time to affect the alloy. Some alloys harden slowly at room temperature and the precipitation can be speeded by heating the alloy within the range 100–200 °C for a given period, termed artificial age-hardening.

Annealing. Annealing softens the alloy and may be used after hardening by cold work or heat treatment. To avoid excessive grain growth the alloy should be heated to the annealing temperature as rapidly as possible and held there only as long as required since grain growth reduces the strength. For work-hardening alloys the temperature range for annealing is 360–425 °C for about 20 minutes, while heat-treatable alloys should be raised to 350–370 °C. Cooling should be in air at not too rapid a rate.

Stabilizing. This consists of heating the alloy to a temperature of about 250 °C and then cooling slowly, the exact temperature depending upon the alloy and its future use. Residual internal stresses are relieved by this treatment.

Classification of aluminium and its alloys

Alloy and heat-treatment designations. Casting alloys (Bs 1490) are prefaced by the letters LM and numbered 1–30, although some have now been withdrawn. Suffixes after the alloy number indicate the condition of the casting thus:

M	as cast
TB	solution heat treated and naturally aged
TB7	solution heat treated and stabilized
TE	artificially aged
TF	solution heat treated and artificially aged
TF7	solution heat treated, artificially aged and stabilized
TS	thermally stress relieved.

Wrought products. These include bar, cold rolled plate, drawn and extruded tube, hot rolled plate, rivet and screw stock, sheet, strip and wire, etc. The mechanical working on the cast metal increases the strength but decreases the ductility. It is in this section that most of the alloys welded occur.

Alloy and temper designations for wrought aluminium (BS 1471 amendment 1 1980). This is an international four digit system of which the first digit indicates the alloy group according to the major alloying elements thus:

1xxx	aluminium of 99.00% minimum purity and higher	5xxx	magnesium
		6xxx	magnesium and silicon
2xxx	copper	7xxx	zinc
3xxx	manganese	8xxx	other elements
4xxx	silicon	9xxx	unused series

The last two digits in group 1xxx show the minimum percentage of aluminium. Thus 1050 is aluminium with a purity (minimum) of 99.50%.

The second digit shows modifications in impurity limits or the addition of alloying elements: the digits 1–9 used consecutively indicate modifications to the alloy. If the second digit is zero, the alloy is unalloyed and has only natural limited impurities.

Four-digit nomenclature referred to previous standard designations

BS and International Designation	Old BS Designation	ISO Designation
1080A	1A	Al 99.8
1050A	1B	Al 99.5
1200	1C	Al 99.0
1350 (electrical purity)	1E	Al 99.5
2011	FC1	Al Cu 5.5 Pb Bi
2014A	H15	Al Cu 4 Si Mg
2031	H12	Al Cu 2 Ni 1 Mg Fe Si
2618A	H16	Al Cu 2 Mg 1.5 Fe 1 Ni 1
3103	N3	Al Mn 1
3105	N31	Al Mn Mg
4043A	N21	Al Si 5
4047A	N2	Al Si 12
5005	N41	Al Mg 1
5056A	N6	Al Mg 5
5083	N8	Al Mg 4.5 Mn
5154A	N5	Al Mg 3.5
5251	N4	Al Mg 2
5454	N51	Al Mg 3 Mn
5554	N52	Al Mg 3 Mn
5556A	N61	Al Mg 5.2 Mn Cr
6061	H20	Al Mg 1 Si Cu
6063	H9	Al Mg Si
6082	H30	Al Si 1 Mg Mn
6101A	91E	Al Mg Si
6463	BTR E6	Al Mg Si
7020	H17	Al Zn 4.5 Mg 1

In groups 2xxx to 8xxx the last two of the four digits only serve to identify the different alloys in the groups. As before the second digit indicates alloy modifications and if it is zero it indicates the original alloy.

There are national variations for some alloys, identified by a letter after the four-numeral designation. The letters are in alphabetical order beginning with A for the first national variation registered, but omitting I, O and Q. An example is 1050A. The accompanying table which refers this BS and international designation with the old BS and ISO designations will make this clear.

Cold work

Temper. This denotes the amount of cold work done on the alloy and has the prefix H followed by (in the UK system) the numbers 1 to 8 indicating increasing strength. The letter O indicates the soft, fully annealed condition and the letter M (F in the US) indicates the material 'as manufactured'.

For example (see table) 3103-0 is an aluminium–manganese alloy in the annealed condition while 3103-H6 indicates the same alloy in the three-quarters hard temper.

The American system has a further figure after the letter H. The figure 1 indicates that the temper was obtained by strain hardening. For example H 16 indicates three-quarters hard temper obtained by strain hardening. The figure 2 indicates that the temper was obtained by strain hardening more than that required and then partially annealing the alloy. The figure 3 indicates that the mechanical properties after cold work are stabilized by a low temperature heat treatment – used only for alloys which otherwise would naturally age-soften at room temperature.

Comparison of UK and US systems

UK symbol	Description	US system
H	Strain hardened, non-heat-treatable material	H
No equivalent	Strain hardened only.	H 1
No equivalent	Strain hardened and partially annealed.	H 2
No equivalent	Strain hardened and stabilized.	H 3
H 2	Quarter hard.	H 12, H 22, H 32
H 4	Half hard.	H 14, H 24, H 34
H 6	Three-quarters hard.	H 16, H 26, H 36
H 8	Fully hard (hardest commercially practicable temper).	H 18, H 28, H 38
No equivalent	A special hard temper (for special applications).	H 19
M	As manufactured.	F
O	Annealed – soft.	O

Symbols used for heat treatment

There are five basic heat-treatment tempers in the UK system and ten basic heat treatment tempers in the US system preceded by the letter T for thermal treatment.

UK symbol	Description	US symbol
—	Cooled from an elevated temperature shaping process and naturally aged to a substantially stable condition.	T1
—	Cooled from an elevated temperature shaping process, cool worked and naturally aged to a substantially stable condition.	T2
T D	Solution heat treated, cold worked and naturally aged to a stable condition.	T3
T B	Solution heat treated and naturally aged to a substantially stable condition.	T4
T E	Cooled from an elevated temperature shaping process and then artificially aged.	T5
T F	Solution heat treated and then artificially aged.	T6
—	Solution heat treated and stabilized.	T7
T H	Solution heat treated, cold worked and then artificially aged.	T8
—	Solution heat treated, artificially aged and then cold worked.	T9
—	Cooled from an elevated temperature shaping process, cold worked and then artificially aged.	T10

Note: Certain other digits are assigned for specific conditions such as stress relieved tempers and can be referred to in the book *Aluminium and its Alloys*, published by The Aluminium Federation (ALFED).

Magnesium

Magnesium is an element of specific gravity 1.8, but although it has a relatively high specific heat capacity (1.1×10^3 joules per kg°C), the volume heat capacity is only $\frac{3}{4}$ of that of aluminium. It melts at 651 °C and its specific latent heat of fusion is lower than that of aluminium, so that for a given section the heat required to melt magnesium is about $\frac{2}{3}$ that required for an equal weight of aluminium. It has a high coefficient of expansion and a high thermal conductivity so that the danger of distortion is always

Percentage composition of typical aluminium alloys for TIG and MIG welding. In each case remainder Al.

Old classification	Alloy designation	Si	Fe	Cu	Mn	Mg	Cr	Zn	
1A	1080A	0.15	0.15	0.03	0.02	0.02	—	0.06	Ga 0.03
1B	1050A	0.25	0.4	0.05	0.05	0.05	—	0.07	
N3	3103	0.5	0.7	0.10	0.9–1.5	0.3	0.1	0.2	Zr + Ti 0.1
N21	4043A	4.5–6.0	0.6	0.3	0.15	0.2	—	0.1	Be 0.0008
N2	4047A	11.0–13.0	0.6	0.3	0.15	0.1	—	0.2	
N6	5056A	0.4	0.5	0.1	0.1–0.6	4.5–5.6	0.2	0.2	Mn + Cr 0.1–0.6
—	5356	0.25	0.4	0.1	0.05–0.2	4.5–5.5	0.05–0.2	0.2	Be 0.0008
N5	5154A	0.5	0.5	0.1	0.1–0.5	3.1–3.9	0.25	0.2	Mn + Cr 0.1–0.5
N4	5251	0.4	0.5	0.15	0.1–0.5	1.7–2.4	0.15	0.15	
N52	5554	0.25	0.4	0.1	0.5–1.0	2.4–3.0	0.05–0.2	0.25	Be 0.0008
N61	5556A	0.25	0.4	0.1	0.6–1.0	5.0–5.5	0.05–0.2	0.2	
H20	6061	0.4–0.8	0.7	0.15–0.4	0.15	0.8–1.2	0.04 0.35	0.25	
H9	6063	0.2–0.6	0.35	0.1	0.1	0.45–0.9	0.1	0.1	

(See also MIG and TIG welding of aluminium alloys, vol. 2.)

Magnesium alloys. Composition (major alloying elements) and properties.

Alloy	Major alloying elements, %	Condition	Tensile st, N/mm²	Elonga-tion, %	Hardness Brinell	Properties
Wrought alloys						
ZTY	Th 0.8, Zn 0.6, Zr 0.6	Extruded forgings	200–230	6–8	50–60	Creep resistant to 350°C. Weldable
ZM21	Zn 2.0, Mn 1.0	Extruded bars, plate, tubes, etc.	220	8–10	50–60	Medium strength sheet and extrusion alloy. Weldable, gas-shielded metal arc.
AZ80	Al 8.5, Zn 0.5, Mn 0.3	Forgings Ppt treated	290	6	60	High strength alloy for forgings of simple design.
AZM	Al 6.0, Zn 1.0, Mn 0.3	Extruded bars, sections, etc.	250	7	55–70	General purpose alloy. Gas and arc weldable.
AZ31	Al 3.0, Zn 1.0, Mn 0.3	Sheet, bars, extruded sections.	220–250	5–12	50–70	Medium strength alloy, sheet, tube, etc. Weldable.
AM503	Mn 1.5	Sheet, plate, extruded sections, etc.	190–230	3–5	35–55	Low strength general purpose alloy, good corrosion resistance. Weldable.
Casting alloys						
ZRE1	Zn 2.5, Zr 0.6, RE 3.0	Ppt. treated, sand cast	140	3	50–60	Creep resistant to 250°C. Pressure tight. Weldable.
RZ5	Zn 4.2, Zr 0.7, RE 1.3	Ppt. treated, sand cast	200	3	55–70	Easily cast, strong at elevated temp.: pressure tight. Weldable
ZE63	Zn 5.8, Zr 0.7, RE 2.5	Soln and Ppt. treated, sand cast	275	5	60–85	Casts well. Pressure tight. Weldable.

ZT1	Th 3.0, Zn 2.2, Zr 0.7	Ppt. treated, sand cast	50–60	5	185	Creep resistant to 350 °C. Pressure tight. Weldable.
TZ6	Zn 5.5, Th 1.8, Zr 0.7	Ppt. treated, sand cast	65–75	5	255	Casts as **RZ5** but stronger. Pressure tight, Weldable.
OH21A	Ag 2.5, Th 1.0, Zr 0.7. Nd rich RE 1.0	Soln and Ppt. treated, sand cast	70–90	2	240	Heat treated alloy, creep resistant: High yield strength to 300 °C. Pressure tight. Weldable.
MSR-B	Ag 2.5, Zr 0.6, Nd rich RE 2.5	Soln and Ppt. treated, sand cast	70–90	2	240	Heat treatable alloy. High yield strength to 250 °C. Pressure tight. Weldable.
A8	Al 8.0, Zn 0.5, Mn 0.3	Soln treated, sand cast	50–60	7	200	Good foundry properties, good ductility and shock resistant.

In addition the following alloys have been introduced:

Wrought ZC61   Zn 6.0%, Cu 1.0%

Casting   QH2LA Zr 0.7%, Ag 2.6%, Th 1.0% Nd 1.0%. Weldable TIG.
EQ21A Zr 0.7%, Ag 1.5%, Nd 2.0%. Weldable TIG.
ZE63A Zr 0.7%, Zn 5.8%, RE 2.5%. Weldable TIG.
Zr55   Zr 0.55%.

Ppt = precipitation, soln = solution, **RE** = rare earths.

present. It oxidizes rapidly in air above its melting point and though it burns with an intense white flame to form magnesium oxide there is little danger of fire during the welding process (TIG and MIG).

As will be seen from the table the chief alloying elements are zinc, aluminium, silver, manganese, zirconium, thorium and the rare earths; the alloying elements improving casting properties, tensile strength, elongation, hardness, etc., as required.

From the table it will be noticed that most of the alloys are weldable by the gas shielded metal arc process. The heat-treatable alloys are solution and precipitation treated in the same way as the aluminium alloys.

As yet standard designations are not equivalent in the British and American Systems, the US having adopted their own Unified Numbering System (UNS) whereas the UK is using the ISO system. The standards applicable are:

USA/ASTM	B80	*Sand castings*
	B90	*Sheet and plate*
	B91	*Forgings*
	B92	*Ingots for remelting*
	B93–94	*Ingots for sand casting, die casting, etc.*
	B1707	*Extruded bars and sections*
UK/BSI	2970	*Ingots and castings*
	3370	*Plate, sheet and strip*
	3372	*Forgings*
	3373	*Extruded bars and sections*
	2901	*Part 4. Filler rods for TIG welding*
	3019	*Part 1. Filler rods for TIG welding*

Stresses and distortion in welding

Stresses set up in welding

In the welding process, whether electric arc or oxy-acetylene, we have a molten pool of metal which consists partly of the parent metal melted or fused from the side of the joint, and partly of the electrode or filler rod.

As welding proceeds this pool travels along and heat is lost by conduction and radiation, resulting in cooling of the joint. The cooling takes place with varying rapidity, depending on many factors such as size of work, quantity of weld metal being deposited, thermal conductivity of the parent metal and the melting point and specific heat of the weld metal.

As the weld proceeds we have areas surrounding the weld in varying conditions of expansion and contraction, and thus a varying set of forces will be set up in the weld and parent metal. When the weld has cooled, these forces which still remain, due to varying conditions of expansion and contraction, are called *residual stresses* and they are not due to any external load but to internal forces.

The stresses will cause a certain deformation of the joint. This deformation can be of two kinds: (1) elastic deformation, or (2) plastic deformation.

If the joint recovers its original shape upon removal of the stresses, it has suffered elastic deformation. If, however, it remains permanently distorted, it has suffered plastic deformation. The process of removal of these residual stresses is termed *stress relieving*.

Stresses are set up in plates and bars during manufacture due to rolling and forging. These stresses may be partly reduced during the process of welding, since the metal will be heated and thus cause some of these stresses to disappear. This may consequently reduce the amount of distortion which would otherwise occur.

Stresses, with their accompanying strains, caused in the welding process, are thus of two types:

(1) Those that occur while the weld is being made but which disappear on cooling.

(2) Those that remain after the weld has cooled.

If the welded plates are free to move, the second causes distortion. If the plates are rigid, the stresses remain as residual stresses. Thus we have to consider two problems: how to prevent distortion, and how to relieve the stress.

Distortion is dependent on many factors, and the following experiment will illustrate this.

Take two steel plates about 150 mm × 35 mm × 8 mm. Deposit a straight layer of weld metal with the arc across one face, using a small current and a small electrode. This will give a narrow built-up layer. No distortion takes place when the plate cools (Fig. 2.30). Now deposit a layer on the second plate in the same way, using a larger size electrode and a heavier current. This will give a wider and deeper layer.

Fig. 2.30

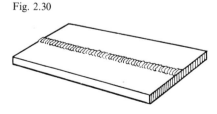

On cooling, the plate distorts and bends upwards with the weld on the inside of the bend (Fig. 2.31). In the first operation the quantity of heat given to the plate was small, due to the small mass of weld laid down. Thus contraction forces were small, and no distortion occurred. In the second operation, much more weld metal was laid down, resulting in a much higher temperature of the weld on the upper side of the plate. On cooling, the upper side contracted therefore more than the lower, and due to the pull of the contracting line of weld metal, distortion occurred. Evidently, if another layer of metal was laid on the second plate in the same way, increased distortion would occur. Thus from experience we find that:

(1) An increase in speed tends to increase distortion because a larger flame (in oxy-acetylene), and a larger diameter electrode and increased current setting (in electric arc) have to be used, increasing the amount of localized heat.

(2) The greater the number of layers of weld metal deposited the greater the distortion.

Let us now consider some typical examples of distortion and practical methods of avoiding it.

Two plates, prepared with a V joint, are welded as shown. On cooling, the plates will be found to have distorted by bending upwards (Fig. 2.32).

Again, suppose one plate is set at right angles to the other and a fillet weld is made as shown. On cooling, it will be found that the plates have pulled over as shown and are no longer at right angles to each other (Fig. 2.33).

This type of distortion is very common and can be prevented in two ways:

(1) Set the plates at a slight angle to each other, in the opposite direction to that in which distortion will occur so that, when cool, the plates are in correct alignment.

Fig. 2.31

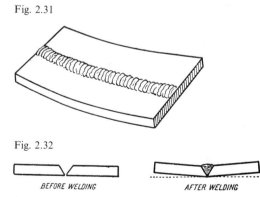

Fig. 2.32

BEFORE WELDING   AFTER WELDING

(2) Clamp the plates firmly in a fixture or vice so as to prevent their movement.

Since we have seen that the amount of distortion depends on several factors, such as speed of welding and number of layers, the amount of bias to be given to the plates in the opposite direction will be purely a matter of experience.

If the plates are fixed in a vice or jig, so as to prevent movement, the weld metal or parent metal must stretch or give, instead of the plates distorting. Thus there is more danger in this case that residual stresses will be set up in the joint.

A very familiar case is the building up of a bar or shaft. Here it is essential to keep the shaft as straight as possible during and after welding so as to reduce machining operations. Evidently, also, neither of the two methods of avoiding distortion given above can be employed. In this case distortion can be reduced to a minimum by first welding a deposit on one side of the shaft, and then turning the shaft through 180 ° and welding a deposit on the opposite diameter. Then weld on two diameters at right angles to these, and so on, as shown in Fig. 2.34. The contraction due to layer 2 will counteract that due to layer 1, layer 4 will counteract layer 3, and so on.

If two flat plates are being butt welded together as shown, after having been set slightly apart to begin with, it is found that the plates will tend to come together as the welding proceeds.

This can be prevented either by tack welding each end before commencing welding operations, clamping the plates in a jig to prevent them moving, or putting a wedge between them to prevent them moving inwards (Fig. 2.35). The disadvantage of tack welding is that it is apt to impair the appearance of the finished weld by producing an irregularity where the weld metal is run over it on finishing the run.

Step welding or back stepping is often used to reduce distortion. In this method the line of welded metal is broken up into short lengths, each length ending where the other began. This has the effect of reducing the heat in any one section of the plate, and it will be seen that in this way when the finish of

Fig. 2.33

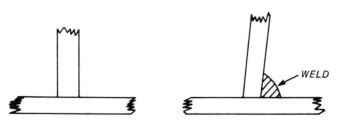

step 2 meets the beginning of step 1 we have an expansion and contraction area next to each other helping to neutralize each other's effect (Fig. 2.36).

In the arc welding of cast iron without pre-heating, especially where good alignment at the end of the welding process is essential, beware of trying to limit the tendency to distort while welding, by tacking too rigidly, as this will frequently result in cracking at the weakest section, often soon after welding has been commenced. Rather set the parts in position so that after welding they have come naturally into line and thus avoid the setting up of internal stresses. Practical experience will enable the operator to decide how to align the parts to achieve this end.

In *skip welding* a short length of weld metal is deposited in one part of the seam, then the next length is done some distance away, keeping the sections as far away from each other as possible, thus localizing the heat (Fig. 2.37). This method is very successful in the arc welding of cast iron.

To avoid distorting during fillet welding the welds can be done in short lengths alternately on either side of the leg of the T , as shown in Fig. 2.38, the welds being either opposite each other as in the sketch (*a*) or alternating as in the sketch (*b*). It is evident that a great deal can be done by the operator to minimize the effects of distortion.

In the case of cast iron, however, still greater care is needed, because whereas when welding ductile metals there is the plasticity of the parent metal and weld metal to cause a certain yielding to any stresses set up, cast

Fig. 2.34

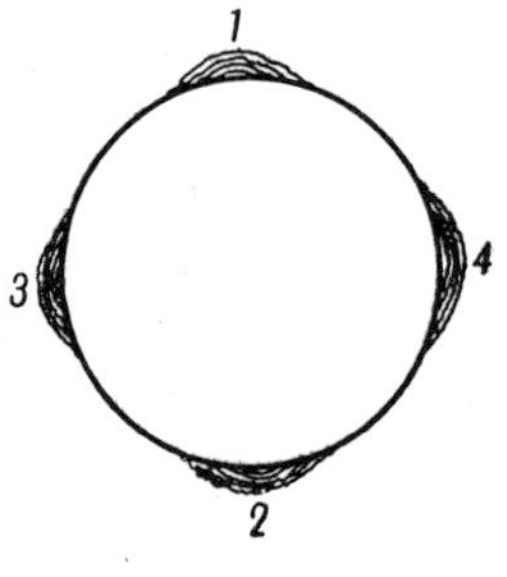

Fig. 2.35

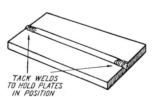

TACK WELDS
TO HOLD PLATES
IN POSITION

iron, because of its lack of ductility, will easily fracture before it will distort, unless the greatest care is taken. This has previously been mentioned in the effects of expansion and contraction.

In the welding of cast iron with the blowpipe, pre-heating is always necessary unless the casting is of the simplest shape. The casting to be welded is placed preferably in a muffle furnace and its temperature raised gradually to red heat.

For smaller jobs pre-heating may be carefully carried out by two or more blowpipes and the casting allowed to cool out in the hot embers of a forge.

Residual stresses and methods of stress relieving

In addition to these precautions, however, the following experiment will show how necessary it is to follow the correct welding procedure to prevent fracture. Three cast-iron plates are placed as shown in Fig. 2.39

Fig. 2.36

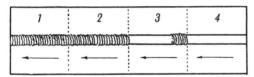

ARROWS AND NUMBERS SHOW DIRECTION AND ORDER OF WELDING

Fig. 2.37

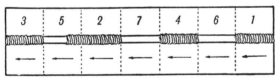

ARROWS AND NUMBERS SHOW DIRECTION AND ORDER OF WELDING

Fig. 2.38

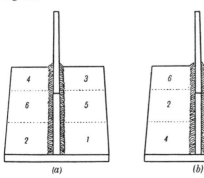

(a)   (b)

and are first welded along the seam *A–B*. No cracking takes place, because they are free to expand and contract.

If we now begin at *D* and weld to *C*, that is, from the free end to the fixed end, cracking will in all probability occur; whereas, if we weld from *C* to *D*, no cracking occurs.

When we weld from *D* to *C* the ends *D* and *C* are rigid and thus there is no freedom in the joint. Stresses are set up in cooling, giving tendency to fracture. Welding from *C* to *D*, that is, from fixed to free end, the plates are able to retain a certain amount of movement regarding each other, as a result of which the stresses set up on cooling are much less and fracture is avoided. Thus, always weld away from a fixed end to a free end in order to reduce residual stress.

Peening. Peening consists of lightly hammering the weld and/or the surrounding parent metal in order to relieve stresses present and to consolidate the structure of the metal. It may be carried out while the weld is still hot or immediately the weld has cooled.

A great deal of controversy exists as to whether peening is advantageous or not. Some engineers advocate it because it reduces the residual stresses, others oppose it because other stresses are set up and the ductility of the weld metal suffers. If done reasonably, however, it undoubtedly is of value in certain instances. For example, in the arc welding of cast iron the risk of fracture is definitely reduced if the short beads of weld metal are lightly peened *immediately* after they have been laid. Care must be taken in peening hot metal that slag particles are not driven under the surface.

Pre- and post-heat treatment

The energy for heat treatment may be provided by oil, propane or natural gas or electricity and the heat may be applied locally, or the welded parts if not too large, may be totally enclosed in a furnace.

The temperature in localized pre-heating, which usually does not exceed 250 °C, should extend for at least 75–80 mm on each side of the welded joint. Post-heat temperatures, usually in the range 590–760 °C, reduce

Fig. 2.39

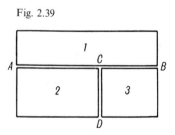

internal stresses and help to soften hardened areas in the HAZ. The work is heated to a given temperature, held at this temperature for a given 'soaking' time, and then allowed to cool, both heating and cooling being subject to a controlled temperature gradient such as 100–200 °C per hour for thicknesses up to 25 mm and slower rates for thicker plate, depending upon the code being followed (ASME, BS, CEGB, etc.).

Localized heat

This can be applied using flexible insulated pad and finger electrical heaters (Fig. 2.40a). These are available in a variety of shapes with

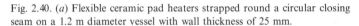

Fig. 2.40. (a) Flexible ceramic pad heaters strapped round a circular closing seam on a 1.2 m diameter vessel with wall thickness of 25 mm.

the elements insulated with ceramic beads and supplied from the welding power source or an auxiliary transformer at 60–80 V.

The pads are connected in parallel as required and are covered over with insulating material to conserve heat. Heat can also be applied locally by gas or oil burners, as for example on circumferential and longitudinal vessel seams (Fig. 2.40*b*). There is now available a rectangular ceramic plaque, protected by an Inconel mesh grill. The plaques each have hundreds of tiny holes and as the gas–air mixture emerges onto the front face of them it is ignited and burns on the plaque surface, which becomes intensely hot (1000 °C), and if these are placed 50–75 mm from the seam to be heated heat is transferred mainly by radiation and rapid heating of the work is achieved

Fig. 2.40. (*b*) Surface combustion units propane or natural gas set up for preheating rotating seams.

up to 250 °C. The burners operate in all positions and can be supplied with magnetic feet for easy attachment to the work.

Furnaces

The older furnaces were of firebrick, fired with gas or oil or electrically heated, and were very heavy. The modern 'top hat' furnace has a hearth with electric heating elements laid in ceramic fibre over which fits a light steel framework insulated inside with mineral wool and ceramic fibre. This furnace is of light thermal mass and can be quickly loaded or unloaded by lifting the unit, consisting of side walls and top, clear of the hearth with its load. The furnace can be heated overnight to take advantage of the lower charges for electrical power, the temperature control of the units being automatic.

These furnaces operate up to 650 °C. If higher temperatures up to 1050 °C are required, additional elements can be added to the furnace sides with loadings up to 1200 kW. Top hat furnaces are made in a variety of sizes including mutiple hearths in which the top hat can be moved onto one hearth while the other is being loaded or unloaded. For furnaces of over 150 m^3 in volume, with their own gas trains, gas firing is very popular. Burners are chassis-mounted and arranged so that they fire down the side walls. As they are of high-velocity gas they avoid direct impingement on the parts being heated and temperatures are evenly distributed by the high-velocity action of the burners.

Temperatures of pre- or post-heat are generally measured by thermo-couple pyrometers and it is very important that the junction of the ends of the thermo-couple wires should be firmly attached to the place where the temperature is to be measured. Clips welded on in the required position and to which the junction is attached are used but when elements are placed over the junction it is possible for a temperature greater than that which actually exists in the work to be indicated due to the proximity of the heating element to the clip joint. A method to avoid this danger is to weld each of the couple wires to the pipe, spaced a maximum of 6 mm apart, by means of a capacitor discharge weld. The weld is easy to make using an auxiliary capacitor unit and reduces the possibility of erroneous readings. The only consumable is the thermo-couple wire, about 150 mm of which is cut off after each operation.

Summary of foregoing section:

Factors affecting distortion and residual stresses

(1) If the expansion which occurs when a metal is heated is resisted, deformation will occur.

(2) When the contraction which occurs on cooling is resisted, a stress is applied.

(3) If this applied stress causes movement, distortion occurs.

(4) If this applied stress causes no movement, it is left as a residual stress.

Methods of reducing distortion

(1) Decrease the welding speed, using the smallest flame (in oxy-acetylene) and the largest diameter electrode and lowest current setting (in electric arc) consistent with correct penetration and fusion of weld and parent metal.

(2) Line up the work to ensure correct alignment on cooling out of the weld.

(3) Use step-back or skip method of welding.

(4) Use wedges or clamps.

Methods of reducing or relieving stress

(1) Weld from fixed end to free end.

(2) Peening.

(3) Heat treatment.

Finally, the following is a summary of the chief factors responsible for setting up residual stresses:

(1) Heat present in the welds depend on:

    (*a*) Flame size and speed in oxy-acetylene welding.

    (*b*) Current and electrode size and speed in arc welding.

(2) Qualities of the parent metal and filler rod or electrode.

(3) Shape and size of weld.

(4) Comparative weight of weld metal and parent metal.

(5) Type of joint and method used in making weld (tacking, back stepping, etc.).

(6) Type of structure and neighbouring joints.

(7) Expansion and contraction (whether free to expand and contract or controlled).

(8) Rate of cooling.

(9) Stresses already present in the parent metal.

Brittle fracture

When a tensile test is performed on a specimen of mild steel it elongates first elastically and then plastically until fracture occurs at the waist which forms. We have had warning of the final failure because the specimen has elongated considerably before it occurs and has behaved both elastically and plastically. It is possible however, under certain conditions, for mild steel (and certain alloy steels) to behave in a completely brittle manner and for fracture to occur without any previous elongation or deformation even when the applied stress is quite low, the loading being well below the elastic limit of the steel.

This type of fracture which takes place without any warning is termed brittle fracture and although it has been known to engineers for many years it was the failure of so many of the all-welded Liberty ships in World War II that focused attention upon it and brought welding as a method of construction into question.

It is not, however, a phenomenon which occurs only in welded fabrications because it has been observed in riveted constructions; but because welded plates are continuous as opposed to the discontinuity of riveted ones a brittle fracture in a welded fabrication may travel throughout the fabrication with disastrous results, whereas in a riveted fabrication it usually travels only to the edge of the plate in which it commenced.

From investigations into the problems which have been made both in the laboratory and on fractures which have occurred during service, it is apparent that certain significant factors contribute to the occurrence of brittle fracture and we may summarize these as follows:

(1) The ambient temperature when failure occurred was generally low, that is, near or below freezing point.

(2) Failure occurred in many cases when the loading on the areas was light.

(3) Failure is generally associated with mild steel but occurs sometimes in alloy steels.

(4) The fracture generally begins at defects such as artificial notches caused by sharp corners, cracks at a rivet hole, a fillet weld corner, poor weld penetration, etc., all behaving generally as a notch.

(5) The age of the structure does not appear to be a significant factor.

(6) Residual stresses may, together with other factors, serve to initiate the fracture.

Once the crack is started it may be propagated as a brittle fracture and can continue at high speeds up to 1200 m per second and at very low loads. It is a trans-granular phenomenon and after failure, tensile tests on other parts of the specimen will show a normal degree of ductility.

We may now consider briefly the conditions which may lead up to the occurrence of brittle fracture.

Notch brittleness

When stress is applied to a bar it may either deform plastically or may break in a brittle manner and the relationship between these types of behaviour determines whether brittle fracture will occur or not.

Consider a specimen of steel in which there is a notch and which is under a tensile stress. The stress on the specimen is below the elastic limit of the main section while at the root of the notch plastic flow may have occurred under a higher stress due to the reduced area.

This localized plastic flow at the notch may cause rapid strain hardening, which may lead to cracking at the root of the notch, and once this crack has begun we have a natural notch of uniform size instead of a variable-size machined notch. Before a crack can be initiated there must therefore be some deformation.

Ductile–brittle transition

The transition from the ductile to the brittle state is affected by temperature, strain rate and the occurrence of notches.

If tests are performed on steel specimens at various temperatures it is found that the yield point stress increases greatly as the temperature is reduced so that the lower the temperature, the greater the brittleness and the liability to brittle fracture. The strain rate has a similar effect on the yield point. As the rate of strain is increased the yield stress rises much more quickly than the fracture stress. A notch may thus localize the stress to increase the strain at its root. The size of the ferrite grains also affects the transition temperature and increasing the carbon content raises and broadens the transition range. In the case of alloy steels, manganese and nickel lower the transition temperature while carbon, silicon and phosphorus raise it. By keeping a high manganese–carbon ratio, tendency to brittle fracture can be reduced and in general the lowest transition temperatures are obtained with finely dispersed microstructures. We may sum up the preceding by saying that the ductile–brittle transition of steel is affected by both grain size and microstructure, and in general the weld metal has similar or even better transition properties than the parent plate. Because of this, brittle fracture is seldom initiated in the weld metal itself but rather in the fusion zone, the heat-affected zone or the parent plate, and the fracture seldom follows the weld. Faults in a weld such as slag inclusion, porosity, lack of fusion, and undercut, which occur in the fusion zone, may serve as nuclei for a crack from which the brittle fracture may be projected, and since the notch ductility of the weld is usually better than that of the plate the fracture follows the plate.

The notch ductility of a weld can be measured by means of the Charpy notch test (p. 269) and gives an indication of the resistance to brittle fracture. The impact values in joules are plotted against the temperature, first for a weld with a low heat input and then with a high heat input, and two curves are obtained as in Fig. 2.41, which are for a carbon–manganese weld metal. It can be seen that the lower heat input gives higher impact values, but there is a transition range from ductile to brittle fracture, and as the temperature falls the probability of brittle fracture is greatly increased. The value of 41 J is often taken as the minimum for the weld metal since above this value it is considered that any crack which develops during service would be arrested before it could result in massive brittle fracture. To obtain good low-temperature impact values, low heat input is therefore required, which means that electrodes should be of the basic type and of small diameter, laid down with stringer bead (split weave) technique. Rutile-coated electrodes can be used down to about $-10\,^\circ$C. Basic mild steel electrodes can be used down to about $-50\,^\circ$C, and below that nickel-containing, basic-coated electrodes should be used.

Fig. 2.41. Ductile–brittle transition in a steel weld metal.

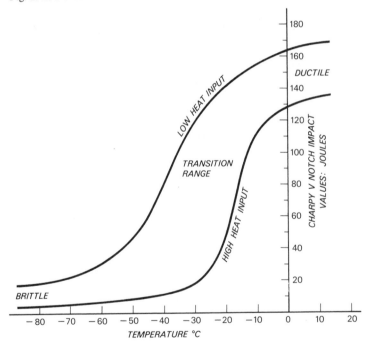

3

Metallic alloys and equilibrium diagrams*

Metallic alloys

The metals with which a welder may have to deal are often not pure metals but alloys consisting of a parent metal with one or more alloying elements added in various proportions, e.g. mild steel is an alloy mainly of iron with a small amount of carbon, while brasses are alloys of copper containing up to 45% of zinc.

The properties of such alloys vary according to the form in which the alloying element is present. There are several possible forms:

(1) The alloying element may be present in an unchanged form in a state of fine mechanical mixture with the parent metal so that it can be seen as a separate particle or crystal (constituent) under the microscope.

Examples of this are:

(a) Carbon present as flakes of graphite in grey cast irons.

(b) Silicon present as fine silicon crystals in the aluminium–silicon alloys.

(c) Lead present as round particles in free cutting brasses. Alloying elements present in this condition do not generally produce a great increase in strength, but they increase or decrease the hardness, reduce the ductility and improve the machinability of the parent metal.

(2) The alloying element may be present in solid solution in the parent metal, i.e., actually dissolved as salt or sugar dissolves in water, so that under the microscope only one constituent, the solid solution, can be seen, similar in appearance to the parent metal except that the colour may be changed.

---

* This chapter will give the welding engineer an introduction to equilibrium diagrams and their uses.

Examples of this are:

(*a*) Up to 37% of zinc can be present in solid solution in copper, causing its colour to change gradually from red to yellow as the percentage of zinc increases. These alloys are known as alpha brasses.

(*b*) Up to about 8% of tin or aluminium can be present in solid solution in copper giving alloys known as alpha tin bronzes or alpha aluminium bronzes respectively.

(*c*) Copper and nickel in any proportion form a solid solution of the same type. Examples are cupro-nickel containing 20–30% nickel and Monel metal containing about 70% nickel. All have a similar appearance under the microscope showing a simple structure similar to that of a pure metal.

Elements in solid solution improve strength and hardness without making the metal brittle.

(3) The alloying element may form a compound with the parent metal which will then have properties different from either. Such compounds are generally hard and brittle and can only be present in small amounts and finely distributed without seriously impairing the properties of the alloy.

These appear under the microscope as distinct new constituents often of clearly crystalline shape. A second type of inter-metallic compound (a new constituent) having ductility and properties more like a solid solution forms in some alloys at certain ranges of composition. An example is the β constituent in brasses containing 40–50% zinc.

(4) While some pairs of metals are soluble in each other in all proportions, e.g. copper and nickel, other pairs can only retain a few per cent of each other in solid solution. If one of the alloying elements is present in a greater amount than the maximum which can be retained in solid solution in the parent metal, then part of it will be in solid solution and part will be present as a second constituent, i.e., either:

(*a*) as crystals of the alloying element which may themselves have a small amount of the parent metal in solid solution, or

(*b*) as an inter-metallic compound.

The state in which an alloying element is present can vary with the temperature so that the constituents present in an alloy at ordinary temperatures may be different from those present at the temperature of the welding operation. For all alloys used in practice, however, the constituents present at any temperature, from that of the welding range down to that of the room, have been determined by quenching the alloys from the various temperatures and examining them under the microscope. Each constituent which can be recognized as separable distinct particles or crystals is called a phase, and an alloy is said to be single-phase, two-phase or three-phase, etc., according to the number of phases present.

For any pair of metals, the different combinations of phrases which may be produced by varying either the composition of the alloy, or the temperature, can be shown in an *equilibrium* or constitutional diagram for that pair of metals. In these diagrams composition is plotted horizontally and temperature vertically. The diagram is subdivided into a number of areas called phase-fields each of which is labelled with the phase or phases which occur within its limits. On the diagram are also plotted curves showing how the melting point and freezing point change with the composition.

Thus a glance at the diagram will show for any composition, the phases present at any chosen temperature. Furthermore the diagram shows what changes will occur in any composition if it is cooled slowly from, say, the welding temperature to normal air temperature.

Equilibrium diagrams and their uses

Two metals soluble in each other in all proportions

In Fig. 3.1 *A* is the melting point of pure copper and *C* that of pure nickel. The percentage composition (Cu and Ni) is plotted horizontally and the temperature vertically.

Fig. 3.1

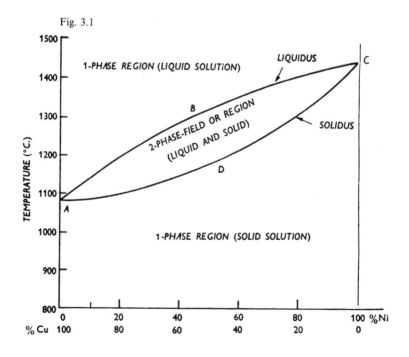

Above the curve *ABC*, which is called the liquidus, any mixture of copper and nickel will be a single liquid solution. This region is called a single-phase field. Below the curve *ADC*, called the solidus, any mixture of copper and nickel will consist of a single solid solution and under the microscope will show no difference between one composition and another except for a gradual change in colour from the red colour of copper to the white of nickel as the percentage of nickel is increased. The region below the curve *ADC* is therefore a single-phase field.

In the region between the curves *ABC*, *ADC* any alloy contains a mixture of:

  (1) solid solution crystals, and
  (2) liquid solution.

This region is therefore a two-phase field. There is a simple rule for determining the compositions and proportions of the liquid and solid solutions which are present together at any selected temperature in a two-phase field. Let *X* be the composition of any selected alloy (Fig. 3.2) and let $t°$ be selected temperature.

Draw a horizontal line at temperature $t°$ to cut the liquidus at *O* and the solidus at *Q*. Then *O* is the composition of the liquid solution (say approx. 28% Ni, 72% Cu) and *Q* the composition of the solid solution (say 73% Ni, 27% Cu) present in the alloy *X* at temperature $t°$. The relative amounts of liquid *O* and the solid *P* are given by the lengths of *PQ* and *OP* respectively.

Fig. 3.2

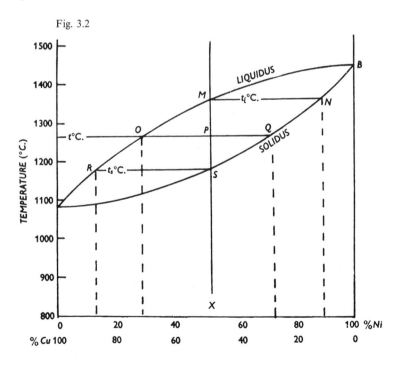

When the alloy of composition X cools from the liquid condition to normal temperatures, the changes involved may be summarized as follows. No change occurs until the alloy cools to the liquidus temperature $t_1°$. When $t_1°$ is reached a small amount of solid solution of composition N (say 90% Ni, 10% Cu) is formed. On further cooling these crystals absorb more copper from the liquid and more crystals separate out but their composition changes progressively along the solidus line towards S getting richer in copper, while the composition of the remaining liquid changes progressively along the liquidus line towards composition R. The amount of solid present therefore increases and the amount of liquid decreases until when the temperature $t_s°$ is reached (this is the solidus temperature of our alloy of composition X) the last drop of liquid solidifies and the whole alloy is then a uniform solid solution of composition X. No further changes occur on cooling to normal temperatures, and under the microscope the cooled alloy will appear to consist of polygonal grains of one kind only, differing from a pure metal only in the colour.

This simplified description applies only when the alloy solidifies very slowly, for time is required for the first crystals deposited (composition N) to absorb copper by diffusion so that they change progressively along NS as solidification occurs. When cooling is rapid there is not sufficient time for these changes to occur in the solidified crystals and we get the first part of each crystal having composition N, so that the centre of each crystal or grain is richer in nickel than the average, while as we progress from the centre to the outside the composition gradually changes, becoming richer in copper until at the boundaries of each grain they are richer in copper than the average composition X. This effect is known as *coring* and is shown as a gradual change in colour of the crystals when the alloy is etched with a suitable chemical solution and examined under the microscope. All solid solution alloys show coring to a greater or lesser degree when they are solidified at normal rates, for example in a casting or welding operation.

Coring can be removed and the composition of each grain made uniform throughout by re-heating or annealing at a temperature just below the solidus of the alloy.

Two metals partially soluble in each other in the solid state, which do not form compounds

This type of alloy system is very common, e.g. copper–silver, lead–tin, aluminium–silicon. It is called a *eutectic system* and at the eutectic composition, that is, at the minimum point on the liquidus curve, the alloy solidifies at a constant temperature instead of over a range of temperatures.

In the solid state there are three phase-fields (Fig. 3.3):

(1) A single-phase region in which any composition consists of a solid solution of metal B in metal A (α only).

(2) A single-phase region in which any composition consists of a solid solution of metal A in metal B (β only).

(3) A two-phase region in which any alloy consists of a mechanical mixture of the two solid solutions mentioned above ($\alpha + \beta$).

Solidification of alloys forming a eutectic system. (1) Alloys containing less than $X\%$ or more than $Z\%$ of metal B will solidify as solid solutions similar to those of the copper–nickel system. In Fig. 3.4 consider an alloy of composition at (1). As it cools, crystals of composition S_1 will begin to separate out at L_1. On further cooling the liquid will change in composition along the liquidus to L_2 and the solid deposited will change along the solidus to S_2. At S_2 the last drop of liquid will solidify and if cooling has been slow, the grains will be a uniform solid solution of composition (1); with rapid cooling the crystal grains will be cored, the centres richer in metal A and the boundaries richer in metal B.

(2) The eutectic alloy will cool as a liquid solution to Y. At Y, crystals of two solid solutions will separate simultaneously, their compositions being

Fig. 3.3

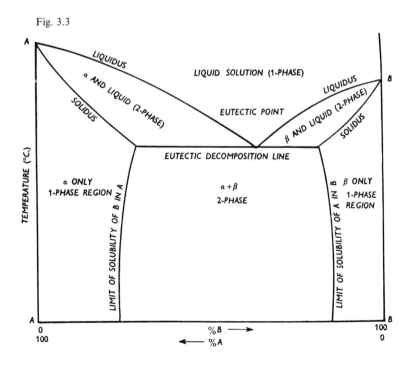

given by the points Z and X. Solidification will be completed at constant temperature and the solidified alloy will consist of a fine mechanical mixture of the two solid solutions, called a eutectic structure.

(3) An alloy of composition between X and Y, or between Y and Z, will begin to solidify as a solid solution but when the composition of the liquid has reached the point Y, the remainder will solidify as a fine eutectic mixture of X and Z.

Two metals which form intermetallic compounds

Compound forming a eutectic system with the parent metal. The equilibrium diagram for many alloying elements which form compounds with the parent metal, e.g. copper alloyed to aluminium, is of a simple eutectic form similar to that just considered for the important part of the diagram, and the changes occurring on cooling an alloy of such a system are similar to those given in (2) above. (In Fig. 3.5, which is for Al–Cu alloy, the intermetallic compound θ formed is $CuAl_2$.)

System in which an intermetallic compound is formed by a reaction. In some alloy systems, notably those in which copper is the parent metal, an intermediate phase or inter-metallic compound is formed as a product of a reaction which occurs in certain compositions during solidification. An example is that of the copper–zinc system in which a β phase is formed by a peritectic reaction.

Fig. 3.4

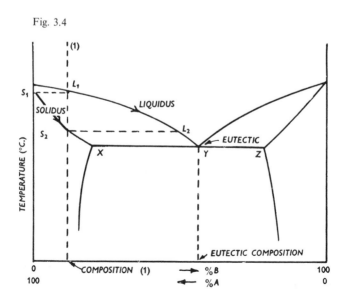

Solidification of alloys in a peritectic system. Referring to Fig. 3.6, alloys containing less than 32.5% of zinc at *A*, solidify as α solid solutions by the process described for class (1). Alloys containing between 32.5% and 39% zinc (*A* and *C* in the figure) commence to solidify by forming crystals of the α solid solution, but when the temperature has fallen to 905 °C, the liquid is of composition at *C* (39% Zn) and the α crystals are of composition at *A*

Fig. 3.5

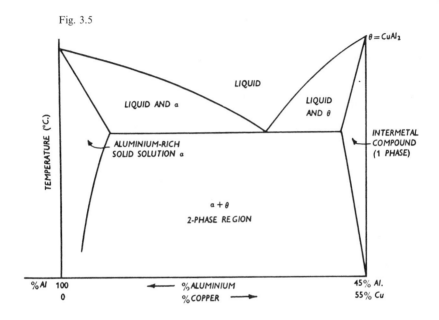

Fig. 3.6

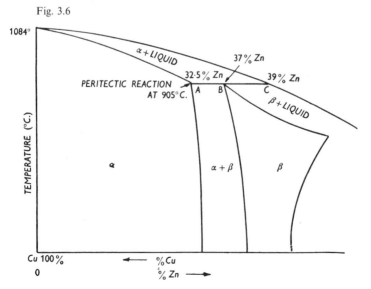

(32.5% Zn). At this temperature the α phase and the liquid react to form a new solution β having a composition at B (37% Zn). This reaction is termed a *peritectic reaction.*

If the mean composition of the alloy is B% zinc (37%) then all the liquid and the α crystals are converted to β solution and at the end of solidification the alloy is a uniform solid solution β.

If the alloy contains between A% and B% of zinc, the alloy after solidification consists of a mixture of α and β. If the alloy contains between B% and C% of zinc, all the α crystals are converted to β at the peritectic reaction, but there is some liquid left which then solidifies as β by the usual mechanism of solidification for a solid solution.

Phase changes in alloys in the solid state

There are two important types of phase change which may occur in alloys while cooling in the *solid* state.

Simple solubility change. In many alloy systems the amount of the alloying element which can be kept in solid solution in the parent metal decreases as the temperature falls.

This effect is shown in the equilibrium diagram Fig. 3.7 by the slope of the solubility line, that is, the line which marks the limit of the single-phase region of the α solid solution.

Thus in the copper–aluminium system, aluminium will dissolve 5.7% of

Fig. 3.7

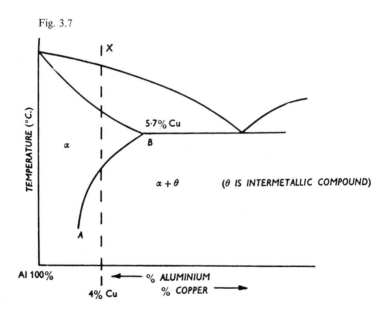

copper at the eutectic temperature but the solubility falls to less than 0.5% copper at ordinary temperatures.

The alloys containing up to 5.7% copper are α solid solutions at temperatures above the line AB, but below the line AB they consist of α solid solutions with varying amounts of an inter-metallic compound θ.

If the alloy of composition X (say 4% copper) is annealed above the line AB, say at 530 °C and quenched in water, a super-saturated solid solution is obtained. If, on the other hand, it is cooled slowly, on reaching the line AB crystals of the inter-metallic compound θ begin to separate out (or are precipitated) and increase in size and number on further cooling.

If a quenched specimen is retained at room temperature, the excess copper in solution tends to diffuse out to form separate θ crystals and this process produces severe hardening of the alloy known as *age hardening*. No visible change in the microstructure occurs.

If a quenched specimen is heated to 100–200 °C the diffusion and hardening occur more quickly. This is known as *temper hardening*, but if the heating is too prolonged, visible crystals of θ separate out and the alloy resoftens. This is known as *over-ageing*.

Eutectoid change. In steels the remarkable changes in properties which can be obtained by different types of heat treatment are the result of a different type of transformation in the solid state. This is known as a *eutectoid transformation*.

Some solid solutions can only exist at high temperatures and on cooling they decompose to form a fine mixture of two other phases. This change, known as a *eutectoid* decomposition, is similar to a eutectic decomposition (see p. 153) except that the solution which decomposes is a solid one instead of a liquid one. The structure produced is also similar in appearance under a microscope to a eutectic structure but is in general finer.

In Fig. 3.8 the β phase undergoes eutectoid decomposition at temperature $t^{1°}$ to form a eutectoid mixture of α phase + γ phase.

If an alloy, which contains some β phase at high temperatures, say composition X in the figure, is quenched from a temperature higher than t_1, the decomposition of β is prevented and severe hardening is produced. If the alloy X is now heated to some temperature below t_1 the change of β to $\alpha + \gamma$ can occur and the hardening is removed. This is then a tempering process.

In some systems in which a change of this type occurs it is possible to obtain a wide range of hardness values in a single alloy according to the rate of cooling from above the decomposition temperature. The *slower* the cooling rate the coarser is the $(\alpha + \gamma)$ eutectoid mixture and the lower the hardness, and vice versa, the quench giving the highest hardness obtainable.

Fig. 3.8

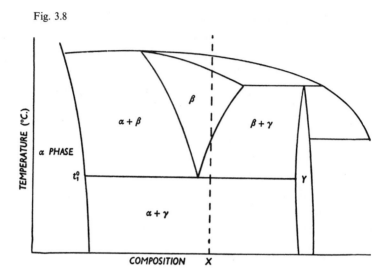

4

Basic electrical principles

Electrical technology

Sources of electrical power
The principle sources of electrical power of interest to the welder are (1) batteries and accumulators, (2) generators.

Batteries generate electrical energy by chemical action. Primary batteries, such as the Leclanché (used for flash lamps and transistor radios), continue giving out an electric current until the chemicals in them have undergone a change, and then no further current can be given out.

Secondary batteries or accumulators are of two types: (1) the lead–acid, and (2) the nickel–iron alkaline. In the former, for example, there are two sets of plates, one set of lead peroxide and the other set of lead, immersed in dilute sulphuric acid (specific gravity 1.250, i.e. 4 parts of distilled water to 1 part of sulphuric acid). Chemical action enables this combination to supply an electric current, and when a current flows from the battery both the lead peroxide plates and the lead plates are changed into lead sulphate, and when this change is complete the battery can give out no more current. By connecting the battery to a source of electric power, however, and passing a current through the battery in the opposite direction from that in which the cell gives out a current, the lead sulphate is changed back to lead peroxide on one set of plates and to lead on the other set. The battery is then said to be 'charged' and is ready to supply current once again.

It may also be noted here that when two different metals are connected together and a conducting liquid such as a weak acid is present, current will flow, since this is now a small primary cell. This effect is called 'electrolysis' and will lead to corrosion at the junction of the metals.

Generators can be made to supply direct current or alternating current as required and are described later. For welding purposes generators of special design are necessary. When welding with alternating current,

159

'transformers' are used to transform or change the pressure of the supply to a pressure suitable for welding purposes.

The electric circuit

The electric circuit can most easily be understood by comparing it to a water circuit. Such a water circuit is shown in Fig. 4.1a. A pump drives or forces the water from the high-pressure side of the pump through pipes to a water meter M, which measures the flow of water in litres per hour. From the meter the water flows to a control valve, the opening in which can be varied, thus regulating the flow of water in the circuit. The water is led back through pipes to the low-pressure side of the pump. We assume that no water is lost in the circuit. Fig. 4.1b shows an electric circuit corresponding to this water circuit. The generator, which supplies direct current (d.c.), that is, current flowing in one direction only, requires energy to be consumed in driving it. The electrical pressure available at the terminals of the generator when no current is flowing in the circuit is known as the electro-motive force (e.m.f.) or in welding as the open circuit voltage. The current is carried by copper wires or cables, which offer very little obstruction or resistance to the flow of current through them, and the current flows from the high-pressure side of the generator (called the positive or +ve pole) through a meter which corresponds to the water meter. This meter, known as an ampere meter or ammeter, measures the flow of current through it in amperes, A (amps for short), just as the water meter measures the flow of water in litres per hour. This ammeter may be connected at any point in the circuit so that the current flows *through* it, since the current is the same at all points in the circuit. From the ammeter the current flows through a copper wire to a piece of apparatus called a 'resistor'. This consists of wire usually made of an alloy, such as manganin,

Fig. 4.1

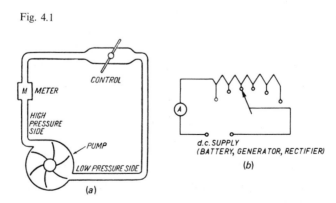

nichrome or eureka, which offers considerable obstruction or resistance to the passage of a current.

The number of turns of this coil in the circuit can be varied by means of a switch, as shown in the figure. This resistor corresponds to the water valve by which the flow of water in the circuit is varied. The more resistance wire which we include in the circuit, the greater is the obstruction to the flow of the current and the less will be the current which will flow, so that as we increase the number of turns or length of resistance wire in the circuit, the reading of the ammeter, indicating the flow of current in the circuit, becomes less. The current finally flows through a further length of copper wire to the low-pressure (negative or − ve) side of the generator.

In the water circuit we can measure the pressure of water in N/mm^2 by means of a pressure gauge. We measure the difference of pressure or potential (p.d.) between any two points in an electric circuit by means of a voltmeter, which indicates the difference of pressure between the two points in volts. Fig. 4.2 shows the method of connexion of an ammeter and a voltmeter in a circuit.

Fall in potential – voltage drop. Let us consider the circuit in Fig. 4.3 in which three coils of resistance wire are connected to each other and to the generator by copper wires as shown, so that the current will flow through each coil in turn. (This is termed connecting them in *series*.) Suppose that the current flows from the + ve terminal *A* through the coils and back to the − ve terminal *H*. Throughout the circuit from *A* to *H* there will be a gradual fall in pressure or potential from the high-pressure side *A* to the low-pressure side *H*. Let us place a voltmeter across each section of the circuit in

Fig. 4.2

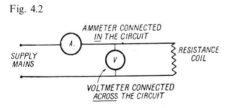

Fig. 4.3

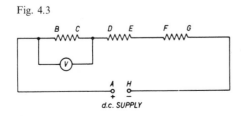

turn and find out where this fall in pressure or *voltage drop* occurs. If the voltmeter is first placed across *A* and *B*, we find that no difference of pressure or voltage drop is registered. This is because the copper wire connecting *A* and *B* offers very little obstruction indeed to the passage of the current, and hence, since there is no resistance to be overcome, there is no drop in pressure.

If the voltmeter is placed across *B* and *C*, however, we find that it will register a definite amount. This is the amount by which the pressure has dropped in forcing the current against the obstruction or resistance of *BC*. Similarly, by connecting the meter across *DE* and *FG* we find that a voltage drop is indicated in each case, whereas if connected across *CD*, *EF*, or *GH*, practically no drop will be recorded, because of the low resistance of the copper wires.

If we add up the voltage drops across *BC*, *DE* and *FG*, we find that it is the same as the reading that will be obtained by placing the voltmeter across *A* and *H*, that is, the sum of the voltage drops in various parts of a circuit equals the pressure applied.

The question of voltage drop in various parts of an electric circuit is important in welding. Fig. 4.4 shows a circuit composed of an ammeter, a resistance coil, and two pieces of carbon rod called electrodes. This circuit is connected to a generator or large supply battery, as shown. When the carbons are touched together, the circuit is completed and a current flows and is indicated on the ammeter. The amount of current flowing will evidently depend on the amount of resistance in the circuit.

If now the carbons are drawn apart about 4 mm, the current still flows across the gap between the carbons in the form of an arc. This is the 'carbon arc', as it is termed. We can control the current flowing across the arc by varying the amount of resistance *R* in the circuit, while if a voltmeter is placed across the arc, as shown, it will register the drop in pressure which occurs, due to the current having to be forced across the resistance of the gap between the electrodes. We also notice that the greater the distance between the electrodes the greater the voltage drop. The metallic arc used in arc welding is very similar to the carbon arc and is discussed fully later.

It has been mentioned that copper wire offers little obstruction or

Fig. 4.4

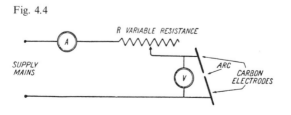

resistance to the passage of a current. All substances offer some resistance to the passage of a current, but some offer more than others. Metals, such as silver, copper and aluminium, offer but little resistance, and when in the form of a bar or wire the resistance that they offer increases with the length of the wire and decreases with the area of cross-section of the wire. Therefore the greater the length of a wire or cable, the greater its resistance; and the smaller the cross-sectional area, the greater its resistance. Thus if we require to keep the voltage drop in a cable down to the lowest value possible as we do in welding, the longer that the cable is, the greater must we make its cross-sectional area. Unfortunately, increasing the cross-sectional area makes the cable much more expensive and increases its weight, so evidently there is a limit to the size of cable which can be economically used for a given purpose.

Series and parallel groupings. We have seen that if resistors are connected together so that the current will flow through each one in turn, they are said to be connected in series (Fig. 4.5*a*). If they are connected so that the current has an alternative path through them, they are said to be connected in parallel or shunt (Fig. 4.5*b*).

An example of the use of a parallel circuit is that of a shunt for an ammeter. The coil in an ammeter has a low resistance and will pass only a small current so that when large currents as used in welding are involved, it is usual to place the ammeter in parallel or shunt with a resistor (termed the ammeter shunt) which is arranged to have a resistance of such value that it carries the bulk of the current, for example 999/1000 of the total current. In this case, if the welding current were 100 A, 99.9 A would pass through the shunt and 0.1 A through the instrument coil, but the ammeter is calibrated with the particular shunt used and would read 100 A (Fig. 4.6*a*).

Fig. 4.5

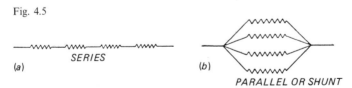

SERIES

(a)

(b)

PARALLEL OR SHUNT

Fig. 4.6. (*a*) Ammeter shunt.

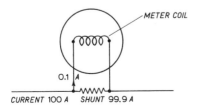

METER COIL

0.1 A

CURRENT 100 A   SHUNT 99.9 A

Link testing ammeter. This useful instrument (Fig. 4.6*b*) enables the current, in either a d.c. or an a.c. circuit, to be measured without breaking the electrical circuit. It is a fluxmeter with pole pieces or links, one of which is fixed and the other moving on a hinge and operated by a trigger on the instrument handle which enables the pole pieces to encircle the circuit with the current to be measured. The indicating meter, calibrated in amperes, is a spring controlled moving iron type in a moulded insulated case. This meter is available in a various range of currents, e.g. 0–50 amps to 0–800 amps, and can be plugged into the instrument body thus increasing its usefulness.

Conductors, insulators and semi-conductors

Substances may be divided into two classes from an electrical point of view: (1) conductors, (2) insulators.

Conductors. These may be further divided into (*a*) good conductors, such as silver, copper and aluminium, which offer very little obstruction to the passage of a current, and (*b*) poor conductors or resistors, which offer quite a considerable obstruction to the passage of a current, the actual amount depending on the particular substance. Iron, for example, offers six times as much obstruction to the passage of a current as copper and is said to have six times the resistance. Alloys, such as manganin, eureka, constantin,

Fig. 4.6 (*b*)

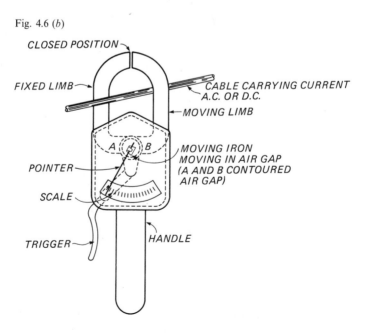

CLOSED POSITION

FIXED LIMB

CABLE CARRYING CURRENT A.C. OR D.C.

MOVING LIMB

A   B

MOVING IRON MOVING IN AIR GAP (A AND B CONTOURED AIR GAP)

POINTER

SCALE

TRIGGER

HANDLE

german silver, no-mag and nichrome, etc., offer much greater obstruction than iron and have been developed for this purpose, being used to control the current in an electric circuit. No-mag, for example, is used for making resistance banks for controlling the current in motor and arc welding circuits, etc., while nichrome is familiar, since it is used as the heating element in electric fires and heating appliances, the resistance offered by it being sufficient to render it red hot when a current flows. Certain rare metals, such as tantalum, osmium and tungsten, offer extremely high obstruction, and if a current passes through even a short length of them in the form of wire they are rendered white hot. These metals are used as filaments in electric-light bulbs, being contained in a bulb exhausted of air, so as to prevent them oxidizing and burning away.

Insulators. Many substances offer such a great obstruction to the passage of a current that no current can pass even when high pressures are applied. These substances are called insulators, but it should be remembered that there is no such thing as a perfect insulator, since all substances will allow a current to pass if a sufficiently high pressure is applied. In welding, however, we are concerned with low voltages. Amongst the best and most familiar insulators are glass, porcelain, rubber, shellac, mica, oiled silk, empire cloth, oils, resins, bitumen, paper, etc. In addition, there is a series of compounds termed synthetic resins (made from phenol and formalin), of which 'bakelite' is a well-known example. These compounds are easily moulded into any desired shape and have excellent insulating properties. Plastics such as polyvinyl chloride (PVC) and chloro-sulphonated-polyethylene (CSP) are now used for cable insulation in place of rubber. PVC is resistant to oil and grease and if ignited does not cause flame spreading. At temperatures near freezing it becomes stiff and more brittle and is liable to crack, while at high temperature it becomes soft. As a result PVC insulation should not be used where it is near a heat source such as an electric fire or soldering iron.

The insulating properties of a substance are greatly dependent on the presence of any moisture (since water will conduct a current at fairly low pressures) and the pressure or voltage applied. If a person is standing on dry boards and touches the + ve terminal of a supply of about 200 volts, the − ve of which is earthed, very little effect is felt. If, however, he is standing on a wet floor, the insulation of his body from the Earth is very much reduced and a severe shock will be felt, due to the much larger current which now passes through his body. As the voltage across an insulator is increased, it is put in a greater state of strain to prevent the current passing, and the danger of breakdown increases.

All electrical apparatus should be kept as dry as possible at all times. Much damage may result from wet or dampness in electrical machines.

Semi-conductors. Most materials are either conductors or insulators of an electric current, but a small group termed semi-conductors fall in between the above types. To compare the resistance of various substances a cube of 1 metre edge is taken as the standard and the resistance between any pair of opposite faces of this cube is termed the resistivity, and is measured in ohm-metre.

Conductors have a low resistivity, copper, for example, being about 1.7×10^{-8} ohm-metre. Insulators have a high resistivity varying between 10^{10} and 10^{18} ohm-metre. Silicon and germanium are semi-conductors and their resistivity depends upon their purity. If a crystal of silicon has a very small amount of antimony or indium added as an impurity its resistivity is lowered, and the greater the amount of impurity the lower the resistivity. If a small amount of antimony is added to one half of a silicon crystal and a small amount of indium to the other half, the junction between these types (termed *n* type and *p* type) acts as a barrier layer, so that the crystal acts as a conductor in one direction and offers a high resistance in the other, in other words it can act as a rectifier. This is the principle of the solid state silicon diode (because it has two elements or connexions) used as a rectifier for d.c. welding supplies.

Welding cables

A cable to conduct an electric current consists of an inner core of copper or aluminium covered with an insulating sheath. Welding cables have to carry quite large currents and must be very flexible and as light as possible, and to give this flexibility the conductor has very many strands of small diameter. The conductor is covered by a thick sheath of tough rubber (TRS) or synthetic rubber (CSP) to give the necessary insulation at the relatively low voltages used in welding. As great flexibility is often not required in the return lead, a cable of the same sectional area but with fewer conductors of larger diameter can be used with a saving in cost.

CSP is a tough flexible synthetic rubber with very good resistance to heat, oils, acids, alkalis, etc., and is a flame retardant. It can be used at higher current densities than a TRS cable of the same sectional area, and its single-sheath construction gives good mechanical strength.

Copper is usually used as the conductor, but aluminium conductors are now used and are lighter than the same sectional area copper, are ecomomical, less liable to pilfering but are larger in diameter and less flexible. In cases where considerable lengths of cable are involved a short length of copper conductor cable can be plugged in between the aluminium cable and electrode holder to give increased flexibility and less liability of conductor fracture at the holder.

Copper-clad aluminium conductors are available with 10% or 20% cladding, increasing the current-carrying capacity for a given cable size and reducing corrosion liability. Clamped or soldered joints can be used and in general TRS cables with copper conductors are probably the best and most economical choice in cases where high-current rating and resistance to corrosive conditions are not of prime importance.

In order to select the correct size of cable for a particular power unit it is customary to indicate for a given cable its current-carrying capacity in amperes (allowing for a permissible rise in temperature) and the voltage drop which will occur in 10 metres length when carrying a current of 100 A, as shown in the table. For greater lengths and currents the voltage drop increases proportionally.

Duty cycle. Cables for welding range from those on automatic machines in which the current is carried almost continuously, to very intermittent manual use in which the cable has time to cool in between load times. To obtain the current rating for intermittently loaded cables the term duty cycle is used. The duty cycle is the ratio of the time for which the cable is carrying the current to the total time, expressed as a percentage. If a cable is used for 6 minutes followed by an off load period of 4 minutes the duty cycle is $6/10 \times 100 = 60\%$. Average duty cycles for various processes are: automatic welding, up to 100%; semi-automatic, 30–85%; manual, 30–60%. Welding cables (BS 638) have many conductors of very small diameter to increase flexibility and may be divided into the following classes:

(1) Single core high conductivity tinned copper (HCC) conductors, paper taped and covered with tough rubber.

(2) Single core HC tinned copper conductors, paper taped and covered with chlorosulphonated polyethylene (CSP)

(3) Single core aluminium conductors (99.5% pure), paper taped and covered with CSP. The CSP cables have a 25% increase in sheath thickness without additional weight.

Ohm's law

Let us now arrange a conductor so that it can be connected to various pressures or voltages from a battery, say 2, 4, 6 and 8 volts, as shown in Fig. 4.7. A voltmeter V is connected so as to read the difference of pressures between the ends of the conductor, and an ammeter is connected so as to read the current flowing in the circuit. Connect the switch first to terminal 1 and read the voltage drop on the voltmeter and the current flowing on the ammeter, and suppose just for example that the readings are 2 volts and 1 ampere. Then place the switch on terminals 2, 3 and 4 in turn and read current and voltage and enter them in a table, as shown below. The last column in the table represents the ratio of the voltage applied to the current flowing, and it will be noted that the ratio is constant, that is, it is the same in each case, any small variations being due to experimental error.

In other words, when the voltage across the conductor was doubled, the current flowing was doubled; when the voltage was trebled, the current was trebled, and so on. This result led the scientist Ohm to formulate his law, thus: The ratio of the steady pressure (or voltage) across the ends of a conductor, to the steady current (or amperes) flowing in the conductor, is constant (provided the temperature remains steady throughout the experi-

TRS and CSP insulated cables, copper conductors

Cross-sectional area of conductor in mm²	Number and diameter of wires. No./mm	Max. overall diameter, mm	Current rating A max. duty cycle						d.c. volts drop 100 A/m of cable at	
			100%		60%		30%		20°C	60°C
			CSP	TRS	CSP	TRS	CSP	TRS		
16	513/0.20	11.5	135	105	175	135	245	190	1.19	1.38
25	783/0.20	13.0	180	135	230	175	330	245	0.78	0.90
35	1107/0.20	14.5	225	170	290	220	410	310	0.55	0.64
50	1566/0.20	17.0	285	220	370	285	520	400	0.39	0.45
70	2214/0.20	19.5	355	270	460	350	650	495	0.28	0.32
95	2997/0.20	22.0	430	330	560	425	790	600	0.20	0.24
120	608/0.50	24.0	500	380	650	490	910	690	0.16	0.18
185	925/0.50	29.0	660	500	850	650	1200	910	0.10	0.12

Fig. 4.7

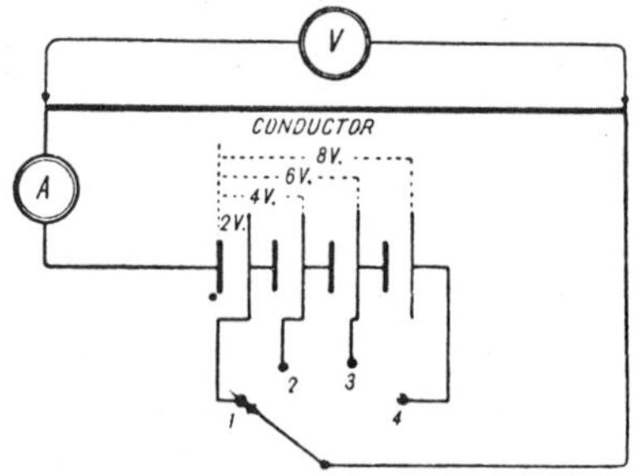

ment, and the conductor does not get hot. If it does, the results vary somewhat. Ohm called this constant the *resistance* of the conductor. In other words, the resistance of a conductor is the ratio of the pressure applied to its ends, to the current flowing in it.

Voltage or pressure drop	Current flowing (amperes)	Ratio: voltage current
2	1	$\frac{2}{1} = 2$
4	2	$\frac{4}{2} = 2$
6	3	$\frac{6}{3} = 2$
8	4	$\frac{8}{4} = 2$

If now a difference of pressure of 1 volt applied to a conductor causes a current of 1 ampere to flow, the resistance of the conductor is said to be 1 ohm, that is, the ohm is the unit of resistance, just as the volt is the unit of pressure and the ampere the unit of current.

That is

$$\frac{1 \text{ volt}}{1 \text{ ampere}} = \text{ohm}$$

or, expressed in general terms,

$$\frac{\text{voltage}}{\text{current}} = \text{resistance.}$$

Another way of expressing this is

$$\text{voltage drop} = \text{current} \times \text{resistance.}$$

A useful way of remembering this is to write down the letters thus: (*V* being the voltage drop, *I* the current, and *R* the resistance). By placing the finger over the unit required, its value in terms of the others is given. For example, if we require the resistance, place the finger over *R* and we find that it equals V/I, while if we require the voltage *V*, by placing the finger over *V* we have that it equals $I = R$. the following typical examples show how Ohm's law is applied to some simple useful calculations.

Example

A pressure of 20 volts is applied across ends of a wire, and a current of 5 amperes flows through it. Find the resistance of the wire in ohms.

$$R = \frac{V}{I} = \frac{20}{5} = 4 \text{ ohms.}$$

Example

A welding resistor has a resistance of 0.1 ohm. Find the voltage drop across it when a current of 150 amperes is flowing through it.

$$V = I \times R,$$
i.e. $V = 150 \times 0.1 = 15$ volts drop.

Power is the rate of doing work, and the work done per second in a circuit where there is a difference of pressure of 1 volt, and a current of 1 ampere is flowing, is 1 *watt*, that is,

power in watts = volts × amperes.

The unit of work, energy and quantity of heat is the joule (J), which is the work done when a force of 1 newton (N) moves through a distance of 1 metre (m). 1 watt (W) = 1 joule per second (J/s). A Newton is that force which, acting on a mass of 1 kilogram (kg), will give it an acceleration of 1 metre per second per second (1 m/s²).

Example

A welding generator has an output of 80 volts, 250 amperes. Find the output in kilowatts and joules per second.

$80 \times 250 = 20\,000$ W
$= 20$ kW
$= 20\,000$ J/s.

This is the actual *output* of the machine. If this generator is to be driven by an engine or electric motor, the power required to drive it would have to be much greater than this, due to frictional and other losses in the machine. A rough estimate of the power required to drive a generator can be obtained by adding on one-half of the output of the generator. For example, in the above,

estimate of power required to drive the generator
$= 20 + 10 = 30$ kW.

It is always advisable to fit an engine which is sufficiently powerful for the work required, and this approximation indicates an engine which would be sufficient for the work including overloads.

Energy is expended when work is done and it is measured by the product of the power in a circuit and the time for which this power is developed. If the power in a circuit is 1 watt for a period of 1 hour, the energy expended in the circuit is 1 watt hour.

The practical unit of energy is 1000 watt hours, or 1 kilowatt hour (kWh), usually termed 1 Unit. This is the unit of electrical energy for consumption purposes, and is the unit on which supply companies base their charge.

Example

An electric motor driving a welding generator is rated at 25 kW. Find the cost of running this on full load, per day of 6 hours, with electrical energy at 2.5p per Unit.

$$\text{Energy consumed in 6 hours} = 25 \times 6 = 150 \text{ kWh or Units.}$$
$$\text{Cost per day} = 150 \times 2.5 = \text{£3.75p.}$$

Resistance of a conductor

The resistance of R ohms of a conductor is proportional to its length l and inversely proportional to its cross sectional area a, that is $R \propto l/a$. Thus the longer a cable the greater its resistance, and the smaller its cross-sectional area the greater its resistance. To reduce the voltage drop in any cable it should be as short as possible and of as large a cross-sectional area as possible.

Measurement of resistance. Ohmmeter and Megger

The Ohmmeter is basically a sensitive current-measuring instrument having its own battery source of supply. When an unknown resistance is connected to the instrument a certain current will flow, causing the pointer to move over the scale. The lower the resistance in circuit, the greater will be the current which flows and consequently the scale, which is graduated in ohms, is in the reverse direction to that of normal instruments, that is zero ohms on the right-hand side of the scale and maximum ohms on the left-hand side. Before taking a reading, the terminals (or the prods attached to the terminals) should be short-circuited and the pointer set to zero ohms with the adjustment provided. If a zero ohms reading cannot be obtained the internal battery should be replaced.

Insulation resistance. The insulation resistance of a cable decreases as its length increases and for lighting and power installations the resistance to earth (that is between any of the conductors and earth) should be not less than 1.5 megohms (one megohm $= 10^6$ ohms). Because of the low voltage of the battery in the Ohmmeter it is unsuitable for measuring the insulation resistance to earth of any installation or appliance since this has normally to be done at twice the installation's normal working voltage, namely 500 volts for a 240 volt installation and 1000 volts for a 450–500 volt installation. This is achieved by using a hand-driven generator incorporated in the instrument. A typical instrument of this type is the 'Megger' which can be used for the measurement of resistance and insulation resistance up to infinity. It has two terminals to which the testing prods are attached and a mechanism ensures that the voltage is limited if the handle is turned too quickly. To perform a test the prods are held one on the conductor and the other on the earth connexion and the handle is rotated. The reading is read directly from the scale graduated similarly to that of the Ohmmeter but ranging from zero ohms to infinity. To test a domestic supply the main switch is placed in the off position and the two outgoing conductors (live and neutral) may be connected together. With all the switches in the off position and one prod on the conductor strap and the other on an earth connexion, the handle is rotated and the reading taken.

Heating effect of a current

When a current flows through a conductor, heat, which is a form of energy, is generated because the conductor has some resistance. The heat generated is proportional to the power in the circuit and the duration for which this power is developed. The power is the product of the volts drop and the current so that:

heat developed $\propto$ power $\times$ time $\propto$ volts drop $\times$ current $\times$ time.

If the current is I amperes flowing for t seconds with V volts drop in a circuit of resistance R ohms, then

heat in joules $= V \times I \times t = I^2 \times R \times t$ (since by Ohm's law $V = I \times R$),

so that the heating effect $\propto$ (current)^2.

Thus if the current in any cable is doubled, four times as much heat is generated in a circuit; if the current is trebled, nine times as much heat is generated. This loss due to the heating effect is known as the $I^2 R$ loss.

The following definitions are useful.

An ampere is that steady current which, passing through two parallel straight, infinitely long conductors of negligible cross-sectional area, one metre apart in a vacuum, produces a force of 2×10^{-7} newtons per metre length on each conductor.

The ohm is that resistance in which a current of 1 ampere flowing for 1 second generates 1 joule of heat energy.

The volt is the potential difference across a resistor having a resistance of 1 ohm and carrying a current of 1 ampere.

Overload protection. One use of the heating effect of a current is to protect electrical apparatus from excessive currents which would cause damage (Fig. 4.8). A heating coil of nickel-chrome wire carries the main current to the apparatus, for example it may be the input supply to a welding transformer. Near the coil is a bi-metal strip made of two thin metal strips, which can be of brass and a nickel alloy rolled together to form a single laminated strip. The brass, which has the greater coefficient of expansion, is placed on the side nearer the heater. When excessive currents flow the position of the strip is such that it curls away with the brass strip on the outer circumference and the contact points break, interrupting the supply to the coil which holds the main contacts in, thus disconnecting the apparatus. The fixed contact can be adjusted to be more or less in contact with the moving contact, thus varying the value of overload current required to break the circuit.

The simple electric circuit of the welding arc

If a metal arc is to be operated from a source of constant pressure, a resistance must be connected in series with it in order to obtain the correct voltage drop across the arc and to control the current flowing in the circuit. This series resistance can be of the variable type, so that the current can be regulated as required. The ammeter A in Fig. 4.9 indicates the current flowing in the circuit, while the voltmeter V_1 reads the supply voltage, and the voltmeter V_2 indicates the voltage drop across the arc. By placing the switch S on various studs, the resistance is varied, and it will be noted that one section of the resistance, marked X, cannot be cut out of circuit. This is to prevent the arc being connected directly across the supply mains. If this happened, since the resistance of the arc is fairly low, an excessive current would flow and the supply mains would be 'short-circuited', and furthermore the arc would not be stable.

The loss of energy in this series resistance is considerable, since a voltage of about 50 to 60 V is required to strike the arc, and then a voltage of about 25 to 30 V is required to maintain it. If then, as in Fig. 4.10, the supply is 60 V and 100 A are flowing in the arc circuit, with 25 volts drop across the arc, this means that there is a voltage drop of 35 V across the resistance. The loss of power in the resistance is therefore (35×100) W $= 3.5$ kW, whereas the power consumed in the arc is (25×100) W $= 2.5$ kW.

Fig. 4.8. Thermal overload trip.

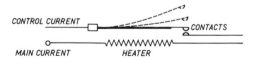

Fig. 4.9

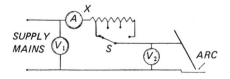

In other words, more power is being lost in the series resistance than is being used in the welding arc. Evidently, therefore, since the 60 V is required to strike the arc, some other more economical means must be found for the supply than one of constant voltage.

Modern welding generators are designed so that there is a high voltage of 50 to 60 V for striking the arc, but once the arc is struck, this voltage falls to that required to maintain the arc, and as a result only a small series resistance is required to control the current, and thus the efficiency of the operation is greatly increased. This type of generator is said to have a 'drooping characteristic'.

This can be illustrated thus: suppose the voltage of the supply is 60 V when no current is flowing, that is, 60 V is available for striking the arc; and suppose that the voltage falls to 37 V when the arc is struck, the voltage drop across the arc again being 25 V and a current of 100 A is flowing (Fig. 4.11). The power lost in the resistance is now only (12 × 100) W or $1\frac{1}{5}$ kW, which is just less than one-half the loss in the previous example, when a constant voltage source was used.

Contact resistance. Whenever poor electrical contact is made between two points the electrical resistance is increased, and there will be a drop in voltage at this point, resulting in heat being developed. If bad contact occurs in a welding circuit, it often results in insufficient voltage being available at the arc. Good contact should always be made between cable lugs and the generator and the work (or bench on which the work rests). The metal plate on the welding bench to which one of the cables from the generator is fixed is often a source of poor contact, especially if it becomes

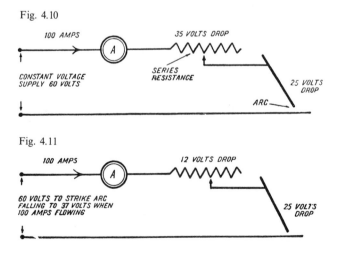

Fig. 4.10

Fig. 4.11

coated with rust or scale. When attaching the return cable to any point on the work being welded, the point should always be scraped clean before connecting the cable lug to it, and in this respect, especially for repair work, a small hand vice, bolted to the cable lug, will enable good contact to be made with the work when the jaws are lightly clamped on any desired point, and this is especially useful when no holes are available in the article to be welded.

Capacitors and capacitance

Principle of the capacitor. Let two metal plates *A* and *B* facing each other a few millimetres apart, be connected through a switch and centre zero milliammeter to a d.c. source of supply (Fig. 4.12*a*). When the switch is closed, electrons flow from *A* to *B* through the circuit, and *A* has a positive and *B* a negative charge. The needle of the meter moves in one direction as the electrons flow and registers zero again when the p.d. between *A* and *B* equals that of the supply. There is no further flow of current and the plates act as a very high resistance in the circuit. The number of electrons transferred is termed the charge, unit charge being the coulomb, which is the charge passing when a current of 1 ampere flows for 1 second.

Between the plates in the air, which is termed the dielectric, there exists a state of electrical stress. If the two plates are now brought quickly together (with the switch still closed) the needle flicks again in the same direction as previously, showing that more electrons have flowed from *A* to *B* and the plates now have a greater charge. When the plates are brought nearer together the positive charge on *A* has a greater neutralizing effect on the charge on *B* so that the p.d. between the plates is lowered and a further flow of electrons takes place. This arrangement of plates separated by a space of dielectric is termed a capacitor and its function is to store a charge of electricity.

Now disconnect the plates from the supply by opening the switch and

Fig. 4.12. (*a*) Charging current, (*b*) discharging current.

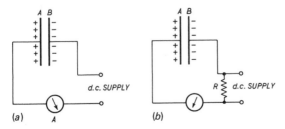

short-circuit the plates through a resistor R (Fig. 4.12b). Electrons flow from B to A, the needle of the meter flicks in the opposite direction, and the charge of electricity which is transferred represents the quantity of electricity which the capacitor will hold and is termed its capacitance. The capacitor is now discharged and in practice the quantity of charge passing is determined by discharging it through a ballistic galvanometer. The moving portion of this instrument has considerable mass and therefore inertia and the angle through which the movement turns is proportional to the quantity of electricity which passes.

Using the same plates as before and about 2 mm apart, charge them through a ballistic galvanometer and note the angle of deflection. Now slide a piece of glass, bakelite, mica or other insulating medium between the plates and after discharging the capacitor repeat the experiment. It will be noted that the angle of deflexion of the meter has increased showing that the capacitance of the capacitor has increased due to the presence of the different dielectric. Similarly if the plates are made larger the capacitance is increased, so that the capacitance depends upon:

(1) The area of the plates. The greater the area, the greater the capacitance.
(2) The distance apart of the plates. The nearer together the plates the greater the capacitance.
(3) The type of dielectric between the plates. Glass, mica and paper give a greater capacitance than air.

Dielectric strength. If the p.d. across the plates of a capacitor is continuously increased, a spark discharge will eventually occur between the plates puncturing the dielectric (if it is a solid). If the dielectric is, say, mica or paper, the hole made by the spark discharge means that at this point there is an air dielectric between the plates, and now it will not stand as high a p.d. across the plates as it did before breakdown.

When capacitors are connected in series (Fig. 4.13b) the sum of the voltage drop across the individual capacitors equals the total volts drop across the circuit. When capacitors are in parallel (Fig. 4.13a) the volts drop across each is the same as that of the supply but the total capacitance is equal to the sum of their individual capacitance.

Capacitance is measured in farads (F). A capacitor has a capacitance of one farad if a charge of one coulomb (C) produces a potential difference of one volt between the plates. This is a very large unit and the sub-multiple is the microfarad (μF). 10^6 μF = 1 farad.

Types of capacitors. The types of capacitors usually met with in welding engineering are the Mansbridge and the electrolytic.

In the Mansbridge type two sheets of thin aluminium foil, usually long and narrow, are separated from each other by a layer of impregnated paper to form the dielectric. Connexions are made to each sheet and the whole is rolled up tightly and placed in an outer container with connexions from the two sheets, one to each terminal. The working voltage and capacitance (in μF) are stamped on the case. These capacitors are rather bulky for their capacitance.

Electrolytic capacitors, on the other hand, have very thin dielectrics and thus can be made in high capacitances but small bulk. If a sheet of aluminium foil is placed in a solution of ammonium borate and glycerine and a current is passed from the sheet to the solution, a thin film of aluminium oxide is formed on the surface of the aluminium. The microscopically thin film is an insulator and insulates the foil from the liquid which is the electrolyte, so that the current quickly falls to zero as the film of oxide forms. The current which continues to flow is the leakage current and is extremely small.

In the wet type of electrolytic, the solution is hermetically sealed in the aluminium canister which contains it, but in the dry type the solution is soaked up in gauze, and the foil, which must be absolutely clean and free from contamination, is rolled up in the gauze, one terminal being connected to the foil and the other to the gauze, and the whole hermetically sealed in an outer case. When connected to a d.c. supply, the initial current which passes forms the dielectric film and the capacitor functions. The

Fig. 4.13. (*a*) Capacitors in parallel. Total capacitance equals the sum of the individual capacitances. (*b*) Capacitors in series. Sum of volts across each equals total drop across the circuit. (*c*) a.c. electrolytic capacitor.

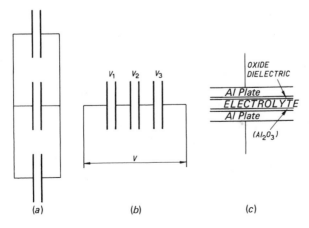

thinner the dielectric, the greater the capacitance, so that for a safe working voltage of say 25 V, the capacitor will be of small bulk even for a capacitance of 1000 μF. The foil terminal is positive so that the capacitor is polarized, and wrong polarity connexions will ruin the unit.

The capacitor just described cannot be used on a.c. with its continuously changing polarity but an a.c. electrolytic capacitor has been developed which, although not continuously rated, is very suitable for circuits where intermittent use is required, as for example the series capacitor which is used to suppress the d.c. component of current when a.c. TIG welding aluminium and its alloys. This type consists of two aluminium electrodes, in foil form separated by gauze soaked in the electrolyte (Fig. 4.13c). When a current flows in either direction, a molecularly thin film of aluminium oxide is formed on each sheet, providing the dielectric. This film is not a perfect insulator and a leakage current flows, which increases with increasing voltage. This leakage current quickly makes good any imperfections in the oxide film but the voltage rating of the unit is critical since excess voltage will produce excess leakage current and lead to breakdown. Capacitors of this type are of small size for capacitances of 1000 μF, and when connected in parallel, are suitable for a.c. circuits in which large currents are flowing, as in welding.

Welding generators

Magnetic field

Pieces of a mineral called lodestone or magnetic oxide of iron possess the power of attracting pieces of iron or steel and were first discovered centuries ago in Asia Minor. If a piece of lodestone is suspended by a thread, it will always come to rest with its ends pointing in a certain direction (north and south); and if it is rubbed on a knitting-needle (hard steel), the needle then acquires the same properties. The needle is then said to be a magnet, and it has been magnetized by the lodestone. Modern magnets of tungsten and cobalt steel are similar, except that they are magnetized by a method which makes them very powerful magnets.

Suppose a magnetized knitting-needle is dipped into some iron filings. It is seen that the filings adhere to the magnet in large tufts near its ends. These places are termed the poles of the magnet. If the magnet is now suspended by a thread so that it can swing freely horizontally, we find that the needle will come to rest with one particular end always pointing northwards. This end is termed the north pole of the magnet, while the other end which points south is termed the south pole.

Let us now suspend two magnets and mark clearly their north and south poles, then bring two north poles or two south poles near each other. We find that they repel each other. If, however, a north pole is brought near a south pole, we find that they attract each other, and from this experiment we have the law: *like poles repel, unlike poles attract* (Fig. 4.14).

If we attempt to magnetize a piece of soft iron (such as a nail), by rubbing it with a magnet, it is found that it will not retain any magnetic properties. For this reason hard steel is used for permanent magnets.

Iron filings provide an excellent means of observing the area over which a magnet exerts its influence. A sheet of paper is placed over a bar magnet and iron filings are sprinkled over the paper, which is then gently tapped. The filings set themselves along definite lines and form a pattern. This pattern is shown in Fig. 4.15.

In the three-dimensional space around the magnet there exists a magnetic field and the iron filings set themselves in line with the direction of action of the force in the field, that is, in the direction of the magnetic flux.

It should be noted that the iron filings map represents the field in one plane only, whereas the flux exists in all directions around the magnet. Fig. 4.16 shows the flux due to two like poles opposite each other and clearly indicates the repulsion effect, while Fig. 4.17 shows the attraction between two unlike poles. The normal flux per unit area (B) is termed the flux density.

Fig. 4.14

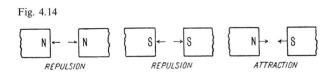

Fig. 4.15. Magnetic field of a bar magnet.

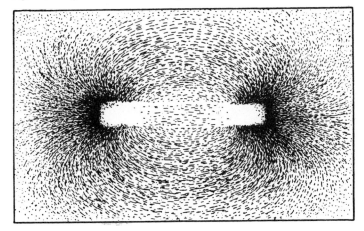

Magnetic field due to a current

If a magnetic needle (or compass needle) is brought near a wire in which a current is flowing, it is noticed that the needle is deflected, indicating that there is a magnetic field around the wire. If the wire carrying the current is passed through the centre of a horizontal piece of paper and an iron filing map made, it can be seen that the magnetic lines of force are in concentric circles around the wire (Fig. 4.18).

Two wires carrying currents in the same direction will attract each other, due to the attraction of the fields, while if the currents are flowing in opposite directions, there is repulsion between the wires.

The magnetic flux round a wire carrying a current is used to magnetize pieces of soft iron to a very high degree, and these are then termed electro-magnets.

Fig. 4.16. Repulsion between like poles.

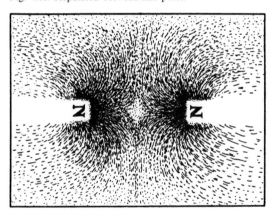

Fig. 4.17. Attraction between unlike poles.

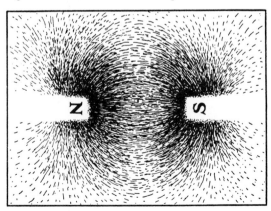

Many turns of insulated wires are wrapped round an iron core and a current passed around the coil thus formed. The iron core becomes strongly magnetized, and we find that the greater the number of turns and the larger the current, the more strongly is the core magnetized. This is true until a point termed 'saturation point' is reached, after which increase in neither the number of turns nor in the current will produce any increase in intensity of the magnetism.

That end of the core around which the current is passing clockwise, when we look at it endways, exhibits south polarity, while the end around which the current passes anti-clockwise exhibits north polarity (Figs. 4.19, 4.20).

Relays and contactors

A solenoid is a multi-turn coil of insulated wire wound uniformly on a cylindrical former and the magnetic flux due to current flowing in a single-turn coil and in a solenoid is shown in Fig. 4.21*a* and *b*. The flux has greatest intensity within the centre of the coil and if a piece of soft iron is

Fig. 4.18

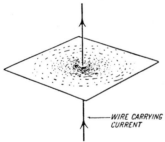

WIRE CARRYING CURRENT

Fig. 4.19

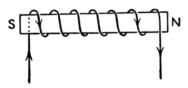

Fig. 4.20

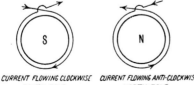

CURRENT FLOWING CLOCKWISE CURRENT FLOWING ANTI-CLOCKWIS
SOUTH POLE   NORTH POLE

placed with one end just inside the coil it becomes magnetized by induction and drawn within the coil when a current flows (Fig. 4.22). In semi-automatic processes such as TIG and MIG, it is important that the various services required, namely, welding current, gas and water, can be controlled from a switch on the welding gun or welding table. This remote-control operation is performed by the use of relays, by which small currents in the control circuit, often at lower voltages than the mains (110 V a.c., 50 V d.c.), operate contactors which make and break the main circuit current.

The control wires to the switch are light and flexible and they control the main welding current, which may be several hundred amperes. Fig. 4.23 shows a simple layout for the control of a welding current circuit. When M is pressed, the control current passes through M and the contactor coil C which is energized, the iron core I moves up, the main contacts are bridged and the main current flows. Also the contacts T are made and because they

Fig. 4.21. (*a*) Magnetic flux due to a single turn of wire carrying a current. (*b*) Magnetic flux due to a coil of wire carrying a current.

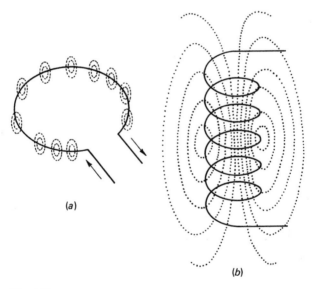

(a)

(b)

Fig. 4.22

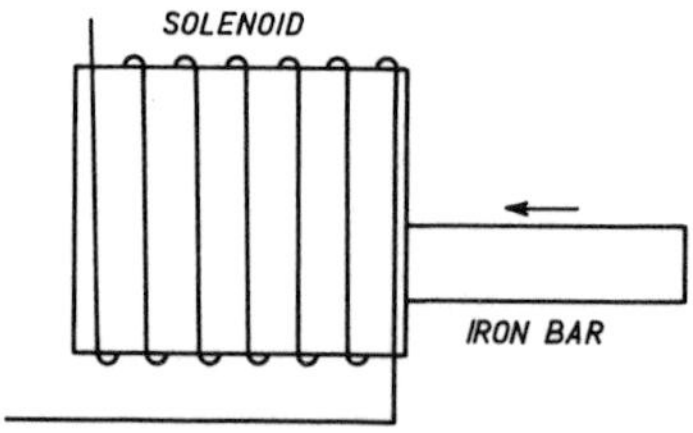

SOLENOID

IRON BAR

are in parallel with those of *M*, when *M* is released, the control current passes through *P*, S_1, S_2 and *T*, keeping the coil energized so that the contactors are held in. Upon pressing *P*, the coil circuit is broken and the contactors break the main circuit when the iron armature falls. Overload heaters are often fitted on two of the phases instead of three because the current in the phases is usually balanced.

The valves are similarly operated by solenoids. When a current flows through the solenoid an armature attached to the valve moves and operates the valve.

Magnetic field of a generator

The magnetic fields of generators and motors are made in this way. The outside casing of the generator, termed the yoke, has bolted to it on its inside the iron cores called *pole pieces*, over which the magnetizing coils, consisting of hundreds of turns of insulated copper wire, fit. To extend the area of influence of the flux, pole shoes are fixed to the pole pieces (or made in one with them), and these help to keep the coils in position. Generators may have 2, 4 or more poles, and the arrangement of a 2- and a 4-pole machine is sketched in Fig. 4.24.

It will be noticed that a north and south pole always come alternately, thus producing a strong flux density where the conductors on the rotating portion of the machine are fixed. The magnetic circuit is completed through the yoke. The coils are connected so that the current passes through them alternately clockwise and anti-clockwise when looked at from the inside of the generator, so as to give the correct polarity (this can be tested by using a

Fig. 4.23. Relay control with overload protection on two phases.

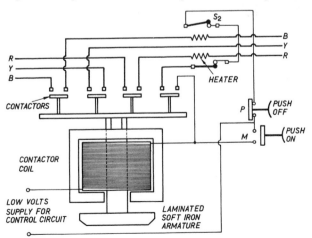

compass needle), and the current which flows through the coils is termed the magnetizing or *excitation* current (Fig. 4.25).

The larger the air gap between the poles of an electro-magnet the stronger must be the magnetizing force be to produce a given flux in the gap. This means that the greater the gap between the pole pieces of a machine and the rotating iron core, called the armature, the greater must the magnetizing current be to produce a field of given strength. For this reason the gap between pole pieces and rotating armature must be kept as small as possible, yet without any danger of slight wear on the bearings causing the armature to foul the pole pieces (the machine is then said to be pole bound). In addition, this gap should be even at each pole piece all round the armature. Excessive air gaps result in an inefficient machine.

The electric field of a d.c. electric motor is similar to that of a generator. Many fractional horse power motors, for example like those used for wire feed in MIG welding, have a magnetic field(s) supplied by permanent magnets. Since these have no windings it greatly simplifies the motor and for FHP motors their reliability and the strength of the permanent magnets makes this method very satisfactory.

Fig. 4.24

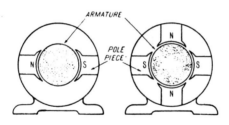

Fig. 4.25

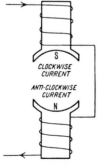

Generation of a current by electrical machines

The following explanation of the principle of operation of a generator is an outline only and will serve to give the operator an idea of the function of the various parts of the machine.

For a current to be generated we require (1) a magnetic flux, (2) a conductor, (3) motion (producing change of magnetic flux).

The magnetic field causes a magnetic flux to be set up, and the conductor is surrounded by this flux. Any change of flux caused by the change in position of the conductor or by change in value of the field will cause a current to be generated in the conductor.

Generation of alternating current

Let us consider the first case. N and S (Fig. 4.26) are the poles of a magnet and AB is a copper wire whose ends are connected to a milliammeter. (This is an instrument that will measure currents of the order of $\frac{1}{1000}$ amp.) If the conductor AB is moved upwards, we find that the needle of A swings in one direction, while if AB is moved downwards, it swings in the opposite direction. By moving AB up and down, we generate a current that flows first in one direction, B to A, and then in the other direction, A to B. This is termed an alternating current, and the current is said to be induced in the conductor.

Note. The rule by which the direction of the current in a conductor is found, when we know the direction of the field and the motion, is termed *Fleming's right-hand rule*. This can be stated thus. 'Place the thumb, first finger and second finger of the right hand all at right angles to each other. Point the first finger in the direction of the flux from N to S and turn the hand so that the thumb points in the direction of motion of the conductor. Then the second finger points in the direction in which the current will flow in the conductor.' Fig. 4.27 makes this clear.

Instead of moving the conductor up and down in this way, the method used for generation is to make the conductor in the form of a coil of several

Fig. 4.26

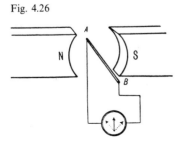

turns and rotate it, and if the coil is wound on an iron core, the field is greatly strengthened and much larger currents are generated.

The ends of the coil are connected to two copper rings mounted on the shaft, but insulated from it, and spring-loaded contacts called brushes bear on these rings, leading the current away from the rotating system (see Fig. 4.28).

From the brushes X and Y, wires lead to the external circuit, which has been shown as a coil of wire, OP, for simplicity.

When the coil is rotated clockwise, AB moves up and CD down. By applying Fleming's rule we find that the current flows from B to A in one conductor and from C to D in the other, as shown by the arrows (Fig. 4.29a). The current will then leave the machine by slip ring Y and flow through the external circuit from O to P and return via slip ring X.

Fig. 4.27. Fleming's right-hand rule.

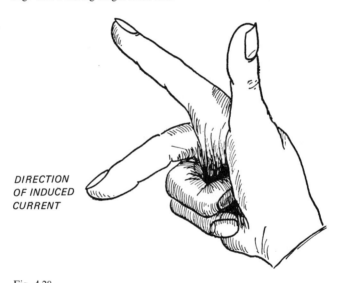

DIRECTION
OF INDUCED
CURRENT

Fig. 4.28

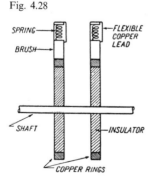

SPRING

BRUSH

←FLEXIBLE
COPPER
LEAD

SHAFT

←INSULATOR

COPPER RINGS

When the coil has been turned through half a turn, as shown in Fig. 4.29*b*, *AB* is now moving down and *CD* up, and by Fleming's rules the currents will now be from *D* to *C* in one conductor and from *A* to *B* in the other. This causes the current to leave by a slip ring *X* and flow through the external circuit from *P* to *O*, returning via slip ring *Y*. If a milliameter with centre zero is placed in the circuit in place of the coil *OP*, the needle of the instrument flicks to one side during the first half turn of the coil and to the other side during the second half turn.

Evidently, therefore, in one revolution of the coil, the direction of the current generated by the coil has been reversed. No current is generated when the coil is passing the position perpendicular to the flux, while maximum current is generated when the coil is passing the position in the plane of the flux. This is illustrated in Fig. 4.30.

Fig. 4.29

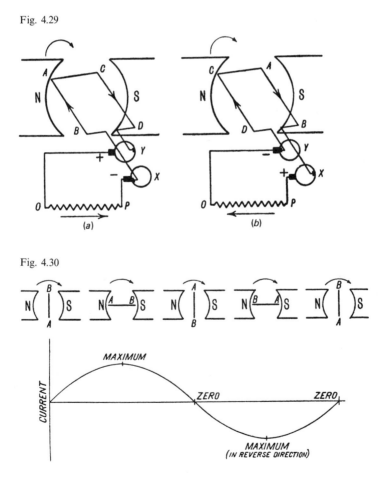

Fig. 4.30

One complete rotation of the coil has resulted in the current starting at zero, rising to a maximum, falling to zero, reversing in direction and rising to a maximum and then falling again to zero. This is termed a complete *cycle*, and the number of times this occurs per second (that is, the number of revolutions which the above coil makes per second) is termed the frequency of the alternating current. 1 cycle per second is known as 1 hertz (Hz), named after the German physicist who discovered electro-magnetic waves.

Alternating currents in this country are usually supplied at 50 Hz. In America 60 Hz is largely used. Evidently a.c. has no definite polarity, that is, first one side and then the other becomes +ve or −ve.

Sinusoidal wave form

The current and voltage generated by a coil rotating in a magnetic field follow a curved path from zero to maximum positions. This curve is known as a sine curve and the voltage and current waves are termed sinusoidal waves.

The e.m.f. generated in a conductor is proportional to the rate at which the conductor cuts the magnetic flux. If the conductor moves across the lines of force so that the flux linkage changes, an e.m.f. is generated. If the conductor moves along a line of force there is no change of flux linkage and no e.m.f. is generated. This can be illustrated in the following way.

The coil AB (Fig. 4.31) is rotating anti-clockwise between the poles N and·S of a magnet and the flux is shown in dotted lines. Let the coil turn from AB to A_1B_1 through an angle θ. From A_1 drop a perpendicular A_1X on to AB and join AA_1. In moving from A to A_1 it can be considered that the conductor has moved from A to X across the flux and from X to A_1

Fig. 4.31

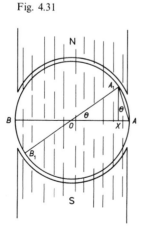

along the flux path. The e.m.f. generated is thus proportional to AX, no e.m.f. being generated in the movement from X to A_1.

But

$$\frac{AX}{AA_1} = \sin \theta,$$

$$\therefore AX = AA_1 \sin \theta$$

and since angle $AA_1X = $ angle AOA_1, the e.m.f. is proportional to the sine of the angle through which the coil is rotated and hence the generated e.m.f. will be a sine curve.

To draw this curve the coil AB (Fig. 4.32) is again rotated anti-clockwise between the magnetic poles N and S. Divide the circle into $30°$ sectors as shown and on the horiontal axis (abscissa) of the graph, divide the $360°$ of one rotation of the coil into $30°$ equal divisions. The vertical axis or ordinate represents the e.m.f. generated. When the coil is in the position AB the conductors are moving along the lines of force and no e.m.f. is generated. As it rotates towards A_1B_1 it begins to cut the lines immediately after leaving the AB position. A_1B_1 is the position of the coil after rotating through $30°$ from AB. Project horizontally from A_1 to meet the vertical ordinate through the $30°$ ordinate at e_1. This ordinate represents the voltage generated at this point. Let the coil rotate a further $30°$ to A_2B_2 and again project horizontally to meet the $60°$ ordinate at e_2. At A_3B_3 the conductors are moving at right angles to the field and generating maximum e.m.f. shown by the e_3 ordinate. Further rotation of the coil results in a reduction of the e.m.f. to zero, e_6, when the coil has rotated through $180°$. After this, further rotation produces a reversal of the e.m.f. and if the points $e_1, e_2, \ldots e_6$ are joined, the resulting curve, is termed as sine wave and represents the generated voltage or e.m.f. Since the current is proportional to the voltage the current wave is also sinusoidal.

Fig. 4.32

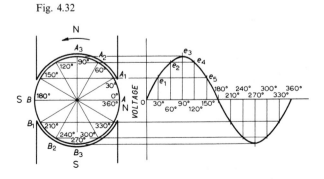

Root mean square value of an alternating current

Since an alternating current or voltage is continuously changing from zero to a maximum value some method must be selected to define the true value of an alternating current or voltage. This is done by comparing the direct current required to produce a given heating effect with the corresponding alternating current which produces the same heating effect.

An alternating current of I amperes is that current which will produce the same heating effect as a direct current of I amperes.

If an alternating current equivalent to I amperes d.c. flows through a resistor $R\Omega$ for t seconds, then the energy generated $= I^2Rt$ joules (p. 172). Let OXY (Fig. 4.33) be the wave form of this current. Divide it into n areas on equal bases, each base being therefore t/n since OY represents the time in seconds. Draw the mid-ordinates $i_1, i_2, i_3 \ldots i_n$ for each area. The energy represented by the first area is $i_1^2R \times t/n$ (mid-ordinate rule for areas).

Energy for second area is $\qquad i_2^2R \times \dfrac{t}{n}$,

Energy for third area is $\qquad i_3^2R \times \dfrac{t}{n}$,

Energy for nth area is $\qquad i_n^2R \times \dfrac{t}{n}$.

Therefore the total energy

$$= i_1^2R \times \frac{t}{n} + i_2^2R \times \frac{t}{n} + i_3^2R \times \frac{t}{n} + \ldots i_n^2R \times \frac{t}{n} \text{ joules}$$

$$= Rt\ \frac{i_1^2}{n} + \frac{i_2^2}{n} + \frac{i_3^2}{n} + \ldots \frac{i_n^2}{n} \text{ joules,}$$

but the total energy is I^2Rt joules, therefore

$$I^2Rt = Rt\ \frac{i_1^2}{n} + \frac{i_2^2}{n} + \frac{i_3^2}{n} + \ldots \frac{i_n^2}{n}$$

$$\therefore \quad I^2 = \frac{i_1^2}{n} + \frac{i_2^2}{n} + \frac{i_3^2}{n} + \ldots \frac{i_n^2}{n}$$

$$\therefore \quad I = \sqrt{\frac{i_1^2 + i_2^2 + i_3^2 + \ldots i_n^2}{n}}$$

That is, the true, effective or virtual value I amperes of the alternating current equals the square root of the mean value of the squares of the current ordinates, or

I = square root of the mean squares, termed the rms value.

If the current wave is sinusoidal this value is 0.707 of the maximum or peak value so that if I_m is the maximum value of the current, the true or rms value is

$$I = 0.707 \ I_m \ (\text{Fig. 4.34}).$$

If an alternating current of maximum value I_m is flowing in a circuit its effect is the same as that of a direct current of value $0.707 \ I_m$.

Similarly with an alternating voltage. If the root mean square (rms) value of a supply is 240 V, the maximum value of the voltage is given by

$$V = 0.707 \ V_m$$
$$240 = 0.707 \ V_m$$
$$V_m = \frac{240}{0.707} = 340 \ \text{V}.$$

This explains why it is possible to get a much greater shock from an a.c. supply of the same rated voltage as a d.c. supply and hence why the earthing of a.c. apparatus is so important. An a.c. supply is always designated by its rms value unless otherwise stated.

Fig. 4.33

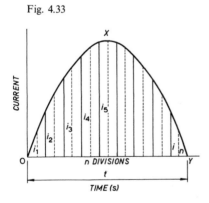

Fig. 4.34

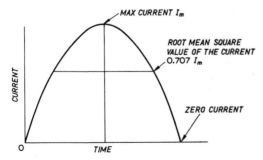

Generation of direct current

By an ingenious yet simple device called the commutator (current reverser), this generated alternating current can be charged to direct current, that is, to a current flowing only in one direction. Instead of slip rings, two segments of copper are mounted on the circumference of the shaft, as shown in Fig. 4.35, being separated from each other by a small gap. Brushes bear on these segments as they did on the slip rings previously. As the coil rotates, the segments will first make contact with each brush in turn and thus reverse the connexions to the external circuit.

In Fig. 4.36*a* and *b*, the conductors are lettered as before, but *AB* is connected to one segment and *CD* to the other. Brushes *X* and *Y* bear on the segments and are connected to the external circuit *OP*. Upon rotating the coil, the current flows in the coil as previously. It leaves by brush *Y* (Fig. 4.36*a*), flows through the external circuit from *P* to *O* and back via brush *X*. In Fig. 4.36*b*, when the coil has turned through half a turn, the connexions of the coils to the brushes have been reversed by the segments of the

Fig. 4.35

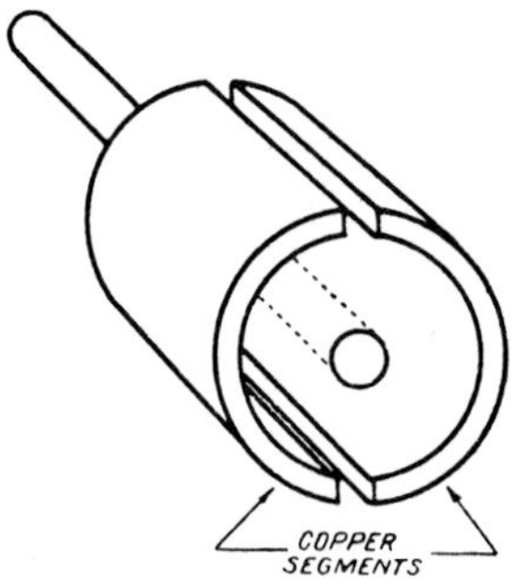

Fig. 4.36

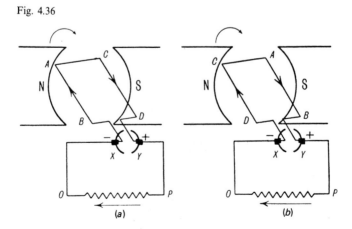

(a)   (b)

commutator and the current again leaves via brush *Y*, through the external circuit in the same way, from *P* to *O*, returning via brush *X*. Thus, though the current in the coil has alternated, the current in the external circuit is uni-directional, or 'direct current' as it is called. Since the current flows from *Y* to *X*, through the external circuit, *Y* is termed the positive pole and *X* the negative pole.

The brushes pass over the joints or gaps between the segments as the coil passes through the position perpendicular to the field, i.e., when no current is generated; thus there is no spark due to the circuit being broken whilst the current is still flowing (Figs. 4.37, 4.38).

The current from a direct current generator with a single coil of several turns, as we have just considered, would be a series of pulsations of current, starting at zero, rising to a maximum and decreasing to zero again, but always flowing in the same direction, as shown in Fig. 4.39.

If, now, a second coil is wound and mounted on the shaft at right angles

Fig. 4.37. Maximum current position.

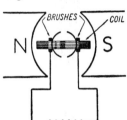

Fig. 4.38. Zero current.

Fig. 4.39

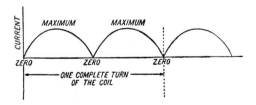

to the first coil and its ends connected to a second pair of commutator segments, the maximum current in one coil will occur when the other coil has zero current; and since there are now four commutator segments, each now only extends round half the length that it did previously. The resulting current from the two coils *A* and *B* will now be represented by a thick line; the dotted portion will no longer be collected by the brushes, because of the shortened length of the commutator segment (Fig. 4.40).

By increasing greatly the number of coils (and consequently the number of commutator segments, since each coil has two segments), the pulsating current can be made less and less, that is, the effect is a steady flow, as shown in Fig. 4.41.

It is not necessary here to enter into details of the various methods of connecting the coils to the segments. Full details of these are given in text-books on electrical engineering. The voltage of a machine is increased by increasing the number of turns of wire in each coil, while the current output of a machine is increased by increasing the total number of coils in parallel on the machine. The output of a machine can also be increased by increasing the speed of the machine and also by increasing the strength of the magnetic flux.

By increasing the number of poles of a machine, its voltage can be increased, while yet keeping its speed the same. Machines of 4 and 6 poles are quite common. In this case there are the same number of sets of brushes as there are poles, i.e. 4 poles, 4 sets of brushes, and so on, and these brushes are connected alternately, as in Fig. 4.42, so as to give +ve and −ve poles.

Fig. 4.40

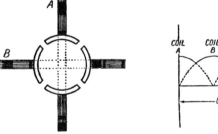

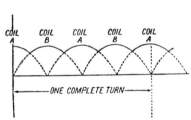

Fig. 4.41

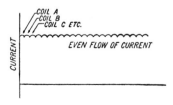

Most welding generators are either 2- or 4-pole. For a given output, the greater the number of poles the slower the speed of the machine. As a rule a machine is designed to operate at a given speed, but the output voltage is varied by a resistor known as the field regulator (see later).

Rectifiers

One method of changing a.c. to d.c. is by the use of an a.c. motor driving a d.c. generator. Static rectifiers perform this operation without the use of moving parts and the types having welding applications are (1) selenium and (2) silicon, and they are used for supplying d.c. for manual metal arc, TIG, MIG, CO_2, etc.

A rectifier should have a low resistance in one direction (forward) so as to allow the current to pass easily, and a high resistance in the opposite direction (reverse) so that very little current will pass. In practice all rectifiers pass some reverse current and this increases with rising temperature so that cooling fins are usually fitted.

Selenium rectifier. This type has largely displaced the copper–copper oxide rectifier for general use and consists of an iron or aluminium base disc coated with selenium, which is a non-metallic element of atomic number 34. A coating of an alloy of lead, cadmium or bismuth is deposited on to the selenium and forms the counter-electrode (Fig. 4.43a).

A current will flow from disc to counter-electrode but not in the reverse direction in which a high resistance is offered, so that the unit acts as a rectifier. To increase the current-carrying capacity the disc is made larger in area and units are connected in parallel. For higher voltages units are connected in series since, if the voltage drop across the unit is too high, the reverse current increases rapidly and the rectifier fails.

Semi-conductor rectifiers. This type is fitted to many of the transformer–rectifier power units for MIG and CO_2 welding. They can

Fig. 4.42. Connexions of brushes.

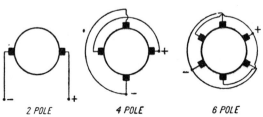

2 POLE   4 POLE   6 POLE

supply large output currents and are generally fan cooled. The following greatly simplified explanation will serve to indicate how they operate.

Silicon is a non-metallic element with four valence electrons in its outer shell. These valence electrons form a covalent bond with electrons in the outer shell of neighbouring atoms by completing the stable octet of eight electrons, which are shared between the two atoms. One way that atoms combine to form a molecule is by means of this bond. The elements neon and argon, for example, have completed outer shells of eight electrons and are therefore completely inert and form no compounds with other elements (Fig. 4.43*b*).

Antimony, phosphorus and arsenic have five electrons in the outer shell, so if a few atoms of antimony are added to a silicon crystal as an impurity (doping), four of the silicon valence electrons form four covalent bonds with four of the antimony valence electrons and there is one free electron due to each antimony atom.

These are termed donor atoms because they can donate an electron, and silicon with this type of impurity is termed n type (negative); if an electron is donated a positively charged ion remains.

Indium, gallium and aluminium have three valence electrons in the outer shell and if, say, indium is added as an impurity to silicon, these three valence electrons form covalent bonds with three of the four silicon valence electrons, but there is one electron missing so that the remaining silicon electron cannot make the fourth covalent bond. This position where the electron is missing is termed a positive hole since it will accept any available electron to form a covalent bond. When the electron enters a positive hole the atom is negatively charged and is a negative ion.

These atoms are termed acceptor atoms because they accept an electron and silicon with this type of impurity is termed p type (positive).

If a silicon crystal is formed, one half being n type and the other half p type, at the junction between the types, electrons will move from n type to p type to fill the holes and the holes will thus move from p type to n type, leaving positive ions in the n type and negative ions in the p type. There is thus a potential barrier of the order of 0.6 V set up across the junction in which there are no free holes or electrons so that electrons tending to move across it are repelled by the negative ions, and holes tending to move are repelled by the positive ions (Fig. 4.43*b*(1)).

If a potential difference is placed across the crystal (Fig. 4.43*b*(2)) so as to make the n type positive and the p type negative, this increases the effect of the potential barrier and no current can flow. This is the reverse direction of the unit.

If the p.d. is reversed, making n type negative and p type positive, the

Fig. 4.43. (*a*) Selenium rectifier.

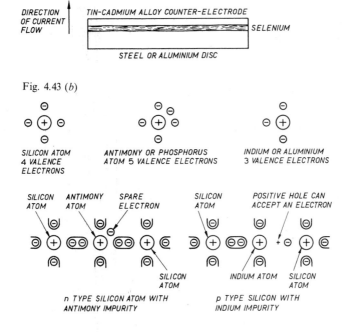

DIRECTION
OF CURRENT
FLOW

TIN-CADMIUM ALLOY COUNTER-ELECTRODE

SELENIUM

STEEL OR ALUMINIUM DISC

Fig. 4.43 (*b*)

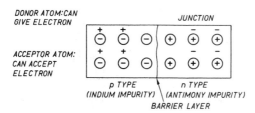

SILICON ATOM
4 VALENCE
ELECTRONS

ANTIMONY OR PHOSPHORUS
ATOM 5 VALENCE ELECTRONS

INDIUM OR ALUMINIUM
3 VALENCE ELECTRONS

SILICON   ANTIMONY   SPARE   SILICON   POSITIVE HOLE CAN
ATOM   ATOM   ELECTRON   ATOM   ACCEPT AN ELECTRON

SILICON
ATOM

INDIUM ATOM   SILICON
               ATOM

n TYPE SILICON ATOM WITH
ANTIMONY IMPURITY

p TYPE SILICON WITH
INDIUM IMPURITY

DONOR ATOM:CAN
GIVE ELECTRON

JUNCTION

ACCEPTOR ATOM:
CAN ACCEPT
ELECTRON

p TYPE
(INDIUM IMPURITY)

n TYPE
(ANTIMONY IMPURITY)

BARRIER LAYER

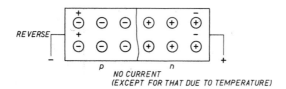

REVERSE

p   *n*

NO CURRENT
(EXCEPT FOR THAT DUE TO TEMPERATURE)

HOLES   ELECTRONS

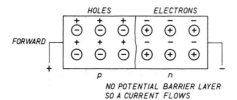

FORWARD

p   *n*

NO POTENTIAL BARRIER LAYER
SO A CURRENT FLOWS

applied p.d. now reduces the potential barrier effect, holes and electrons move and current flows. This is the forward direction of the current. A very small leakage current always occurs at normal temperatures due to the breaking of some of the covalent bonds but as the temperature of the junction rises, breaking of bonds increases, the carriers across the junction accelerate and remove other bonds and there comes a point when the junction breaks down and a reverse or breakdown current flows (Fig. 4.44a). The conductivity of both p and n type silicon is increased over pure silicon depending on the amount of doping; because the unit has two connections it is referred to as a diode. Only one half of the a.c. wave flows, the other half being suppressed; this is termed half-wave rectification and the current and voltage consist of uni-directional pulses of 50 per second in the case of a 50 Hz supply. To obtain full-wave rectification the rectifier elements are connected in 'bridge' connection as in Fig. 4.44b.

Fig. 4.44 (a). The diode.

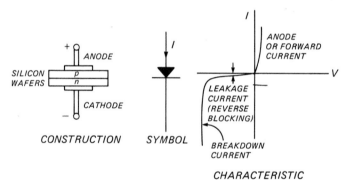

Fig. 4.44 (b). Single-phase full-wave rectification.

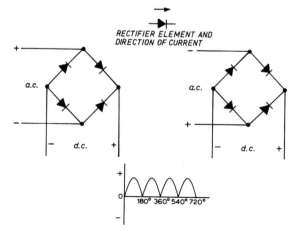

If full-wave rectification on a three-phase system is required the elements are connected as in Fig. 4.44c, the output of the secondary of the transformer being delta connected. This method is favoured in many cases for d.c. welding units and gives a balanced load.

Thyristors

A thyristor is a solid state switch which consumes very little power and is small and compact. Thyristors can be used for a variety of switching operations, as for example in motor control, resistance welding, circuit control and control of welding power sources. Large thyristors can carry heavy currents of up to several hundred amperes.

It is similar in construction to a solid state diode but has four elements of doped silicon in alternate layers p n p n and a gate connexion to the p element on the cathode side (Fig, 4.45a). If there is no connexion to the gate terminal the thyristor behaves as three diodes in series, pn np pn, so that current in either direction is blocked and with the cathode + ve (reverse) it is similar to the diode (Fig. 4.44a).

Fig. 4.44 (c)

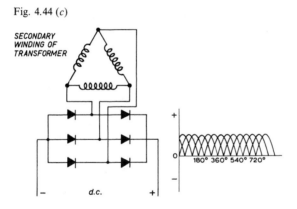

Fig. 4.45 (a). The thyristor.

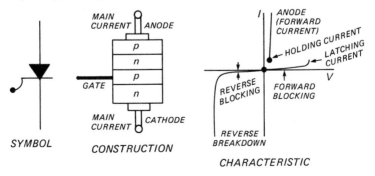

When the anode is made + ve (forward), only a leakage current will flow until a breakdown voltage across the pn junction is reached, when almost all the voltage appears across the junction next to the cathode.

To reach the 'on' state, the thyristor, anode + ve, must attain a 'latching level', this being quite a low level of the full load value and can be obtained by putting in a gate current, forming holes in the p wafer which, with electrons from the n wafer next to the cathode, causes breakdown of the control junction next to the cathode. The anode current is now over latching level, the thyristor is now switched 'on' and in this state no further gate current is required, the thyristor remaining 'on'.

By varying the voltage bias on the gate, as for example with a potentiometer, stepless control of current is obtained from the thyristor.

Since the thyristor suppresses one half of the a.c. wave it is necessary to connect the thyristors with diodes in bridge connexion to obtain full-wave rectification as when a stepless d.c. power output is required in a welding source. This is shown in Fig. 4.45b. The value of welding current is set by the single knob current control C on the front panel. This value is compared by the control circuits with the value of output current received from H and the firing angle of the thyristors is altered to bring the output current to that value set by the current control.

Ignitron

The ignitron is a rectifier which has three electrodes: an anode, a cathode which is a mercury pool and an igniter, all contained in an evacuated water-cooled steel shroud. The igniter is immersed in the mercury cathode, and when a current is passed through it a 'hot spot' is formed on the surface of the pool. This acts as a source of electrons which stream to the anode if it is kept at a positive potential with respect to the

Fig. 4.45 (b). Thyristor control of welding current. Simplified schematic diagram for one knob control of welding current using thyristors or silicon controlled rectifiers (SCR).

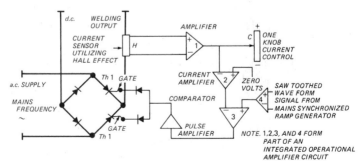

cathode, and a current can now flow from anode to cathode through the ignitron. It will continue doing so as long as the anode remains positive, and large ignitrons can handle currents of several thousands of amperes. The single ignitron behaves as a half-wave rectifier on an a.c. supply, and two connected in reverse parallel can operate as a switch controlling the flow of a.c. in a circuit. The arc within the unit can be struck at any point in any particular half cycle by controlling the current in the igniter circuit. The ignitron has been mostly replaced by the thyristor.

The inverter

The inverter is an electrical device which converts d.c. to a.c., that is, it is the opposite to a rectifier.

In any welding unit the greatest weight is that of the transformer, single- or three-phase, because of the weight of the closed-circuit laminated steel magnetizable core.

As the frequency of an a.c. increases, the size and thus the weight of this laminated core becomes less because the inductance of the circuit increases with the frequency, so that if we can feed a high-frequency a.c. into a transformer so as to get a welding output voltage, this transformer will be very much less in weight than the conventional type at 50 Hz because of the reduced iron circuit.

In the illustration, Fig. 4.45c, a three-phase 50 Hz supply is passed into an inverter which converts it to a.c. and increases its frequency to thousands of Hz. This high-frequency (HF) a.c. is passed into a transformer which has a magnetic circuit only a fraction of that in a conventional type and it is stepped down to a suitable welding supply voltage before being passed into a bridge rectifier, the output from which is the d.c. welding supply. A unit, for example, of about 400 A output weighs about 50 kg and is thus much lighter than the conventional unit.

The control system is connected between inverter and output rectifier and a smooth arc with good starting characteristics results.

This light weight, with high output, makes it possible to transport it by car and it can be taken and used, near to the operator, on sites which may

Fig. 4.45 (c)

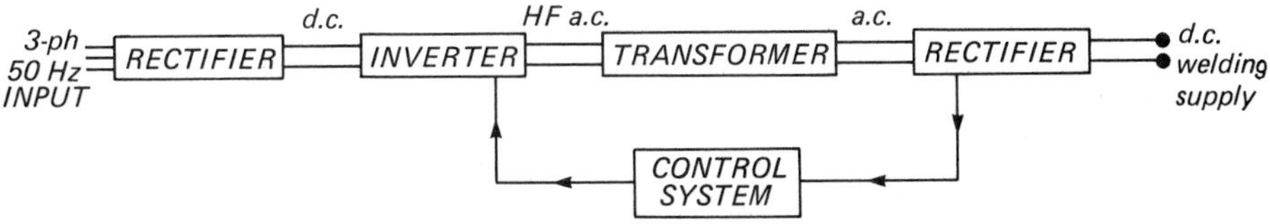

otherwise need long welding cables. Longer mains cables can be used and these are lighter and more convenient.

Description of a typical direct-current generator

A modern direct-current welding generator consists of:

(1) Yoke with pole pieces and terminal box, and end plates.
(2) Magnetizing coils.
(3) Armature and commutator (the rotating portion).
(4) The brush gear.

The yokes of modern machines are now usually made of steel plate rolled to circular form and then butt welded at the joint. The end plates, which contain the bearing housings, bolt on to the yoke, and the feet of the machine are welded on and strengthened with fillets. The pole pieces are of special highly magnetizable iron and are bolted onto the yoke. The coils are usually of double cotton-covered copper wire, insulated and taped overall, and they fit over the pole pieces, being kept in position by the pole shoes. The armature shaft is of nickel steel and the armature core (and often the pole pieces also) is built up of these sheets of laminations of highly magnetizable iron, known by trade names such as Lohys, Hi-mag, etc. Each lamination is coated with insulating varnish, and they are then placed together and keyed on to the armature shaft, being compressed tightly together so that they look like one solid piece.

The insulating of these laminations from each other prevents currents (called eddy currents) which are generated in the iron of the armature when it is rotating from circulating throughout the armature and thus heating it up. This method of construction contributes greatly to the efficiency and cool running of a modern machine. The armature laminations have slots in them into which the armature coils of insulated copper wire are placed (usually in a mica or empire cloth insulation). The coils may be keyed into the slots by fibre wedges and the ends of the coils are securely soldered (or sweated) on to their respective commutator bars (the parts to which they are soldered are known as the commutator risers). A fan for cooling purposes is also keyed on to the armature shaft.

The commutator is of high conductivity, hard drawn copper secured by V rings, and the segments are insulated from the shaft and from each other by highest quality ruby mica. Brushes are of copper carbon, sliding freely in brush holders, and springs keep them in contact with the commutator. The tension of the springs should only be sufficient to prevent sparking. Excessive spring pressure should be avoided, as it tends to wear the commutator unduly. The commutator and brush gear should be kept clean by occasional application of petrol on a rag, which will wash away accumulations of carbon and copper dust from the commutator micas and brush gear. All petrol must evaporate before the machine is started up, to avoid fire risk. The armature usually revolves on dust-proof and watertight ball or roller bearings, which only need packing with grease every few months. Older machines have simple bronze or white metal bearings, lubricated on the ring oil system. These need periodical inspection to see that the oil is up to level and that the oil rings are turning freely and, thus, correctly lubricating the shaft.

Connexions from the coils and brush gear are taken to the terminal box of the machine, and many welding generators have the controlling resistances and meters also mounted on the machine itself.

Connexions of welding generators

In the following sketches, magnetizing coils are shown thus: —eeeeeee—, and this represents however many coils the machine possesses, connected so as to form alternate north and south poles, as before explained. The armature, with the brushes bearing on the commutator, is shown in fig. 4.46a.

The current necessary for magnetizing the generator is either taken from the main generator terminals, when the machine is said to be self-excited. Welding generators are manufactured using either of these methods.

Separately excited machines

These generally take their excitation or magnetizing current from a small separate generator, mounted on an extension of the main armature shaft, and this little generator is known as the exciter. Current generated by this exciter passes through a variable resistor, with which the operator can control the magnetization current, and then round the magnetizing coils of the generator. This is shown in Figs. 4.46b and 4.51.

By variation of the resistance R, the magnetizing current and hence the strength of the magnetic flux can be varied. This varies the voltage (or pressure) of the machine and thus enables various voltages to be obtained

Fig. 4.46 (*a*)

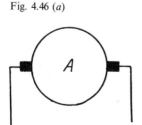

across the arc, varying its controllability and penetration. This control is of great importance to the welder.

This type of machine gives an almost constant output voltage, irrespective of load, and thus, as before explained, would result in large losses in the series resistor, if used for welding. In order to obtain the 'drooping characteristic', so suitable for welding, the output current is carried around some *series* turns of thick copper wire, wound over the magnetizing coils on the pole pieces, and thus current passes round these turns so as to magnetize them with the opposite polarity from the normal excitation current. Fig. 4.47 shows how the coils are arranged. Consider then what happens.

When no load is on the machine, the flux is supplied from the separate exciter and the open circuit voltage of the machine is high, say 60 volts, giving a good voltage for striking. When the arc is struck, current passes through the series winding and magnetizes the poles in the opposite way from the main flux and thus the strength of the flux is reduced and the voltage of the machine drops. The larger the output current the more will the voltage drop, and evidently the voltage drop for any given output current will depend on the number of series turns. This is carefully arranged

Fig. 4.46 (*b*). Simple separately excited generator.

Fig. 4.47

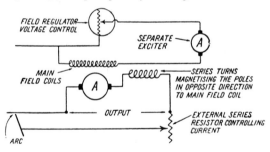

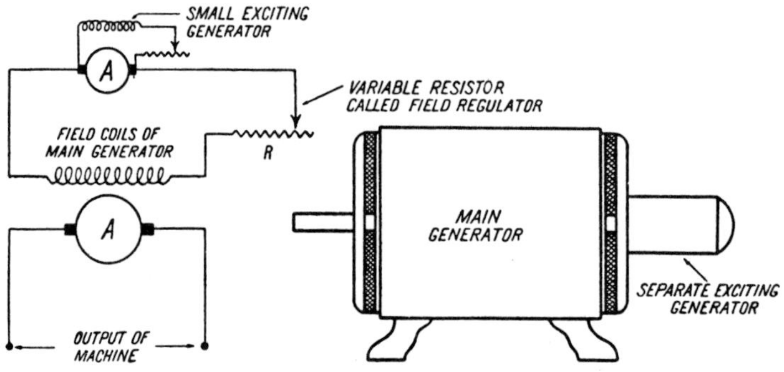

when the machine is manufactured, so as to be the most suitable for welding purposes.

This type of machine, with control of both current and voltage, is very popular and is reliable, efficient and economical. Because the voltage available at any given instant is only slightly greater than that required to maintain the arc, only a small series resistance is required, this being fitted with the usual variable control.

Self-excited machines

The simplest form of this type of machine is that known as the 'shunt' machine, in which the magnetizing coils take their current direct from the main terminals of the generator through a field-regulating resistor (Fig. 4.48).

There is always a small amount of 'residual' magnetism remaining in the pole pieces, even when no current is passing around the coils, and, when the armature is rotated, a small voltage is generated and this causes a current to pass around the coils, increasing the strength of the flux and again causing a greater e.m.f. to be generated, until the voltage of the machine quickly rises to normal. Control of voltage is made, as before, by the field regulator. This type of machine is not used for welding because its voltage only drops gradually as the load increases and, as before explained, this would cause a waste of energy in the external series resistor.

Again, this machine is modified for use as a welding generator by passing the output current first round series turns wound on the pole pieces, so as to magnetize them with the opposite polarity from that due to the main flux, and this results, as before, in the voltage dropping to a great extent as the load increases and, thus, the loss of energy in the external resistor is greatly reduced (Fig. 4.49). A machine of this type is termed a differential compound machine and shares with the separately excited machine the distinction of being a reliable, efficient and economical generator for welding purposes. The control of current and voltage are exactly as before.

The rest of the equipment of a direct current welding generator consists of a main switch and fuses, ammeter and voltmeter. The fuses have an

Fig. 4.48. Connexions of a simple shunt machine.

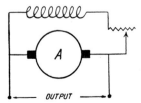

insulating body with copper contacts, across which a piece of copper wire tinned to prevent oxidation is bridged. The size of this wire is chosen so that it will melt or 'fuse' when current over a certain value flows through it. In this way it serves as a protection for the generator against excessive currents, should a fault develop.

On many machines neither switch nor fuses are fitted. Since there is always some part of the external resistor connected permanently in the circuit of these machines, no damage can result from short circuits, and fuses are therefore unnecessary. The switch is also a matter of convenience and serves to isolate the machine from the electrode holder and work when required.

Interpoles, or commutation poles

Interpoles are small poles situated between the main poles of a generator and serve to prevent sparking at the brushes. The polarity of each interpole must be the same as that of the next main pole in the direction of rotation of the armature, as in Fig. 4.50a.

They carry the main armature current and, therefore, like the series winding on welding generators, are usually of heavy copper wire or strip.

They prevent distortion of the main flux, by the flux caused by the current flowing in the armature, and thus communication is greatly assisted.

Most modern machines are fitted with interpoles, as they represent the

Fig. 4.49

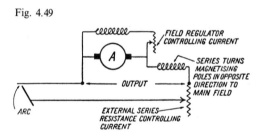

Fig. 4.50 (a)

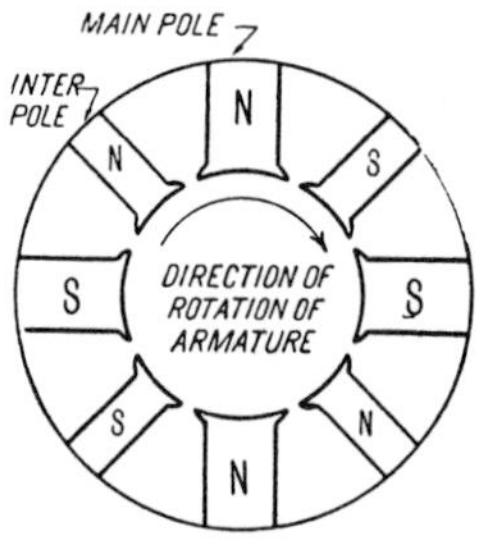

most convenient and best method of obtaining sparkless commutation.

The following is a summary of the features of a good welding generator:

(1) Fine control of voltage.
(2) Fine control of current.
(3) Excitation must always provide a good welding voltage.
(4) Copper conductors of armature and field of ample size and robust construction, yet the generator must not be of excessive weight.
(5) Well-designed laminated magnetic circuit and accurate armature-pole shoe air gap.
(6) Good ventilation to ensure cool running.
(7) Well-designed ball or roller bearings of ample size and easily filled grease cups.
(8) Well-designed brush gear – no sparking at any load and long-life brushes.
(9) Bearings and brush gear easily accessible.
(10) Large, easily placed terminals enabling polarity to be quickly changed (or fitted with polarity changing switch).
(11) High efficiency, that is, high ratio of output to input energy. (60–65% efficiency is normal for a modern single-operator motor-driven direct current plant.)

Brushless alternators and generators

Brushless alternators and generators have no slip rings or commutators but use a rectifier mounted on the rotating unit (rotor) to supply direct current to excite the rotating field coils, the main current being generated in the stationary (stator) coils.

The rotor has two windings, an exciter winding and a main field winding. The exciter winding rotates in a field provided by exciter coils on the stator and generates a.c., which is passed into silicon diodes (printed circuit connected) mounted on the rotor shaft. The resultant d.c. passes through the rotating field coils of the main generator portion and the rotating field produced generates a.c. in the stationary windings of the main generator portion. If a d.c. output is required, as for welding purposes, this a.c. is fed into a silicon rectifier giving a d.c. output (Fig. 4.50*b*).

Current for the stator coils on the exciter is obtained from the rectifier supply and variation of the excitation current gives variable voltage control as on a normal generator, and residual magnetism causes the usual build-up.

Dual continuous control generator

In the dual continuous control generator, excitation current is supplied by the separate exciting generator shown on the left of Fig. 4.51*a*, and the control of the excitation current is made by the field rheostat, which therefore controls the output voltage of the machine. Interpoles are fitted to prevent sparking and the continuously variable current control is in parallel with the differential series field, the current control being wound on a laminated iron core so as to give a smoothed output. This generator gives a good arc with excellent control over the whole range and is suitable for all classes of work.

Generators in parallel

The parallel operation of generators enables the full output of the machines to be fed to a single operator. If two shunt wound generators are

Fig. 4.50 (*b*)

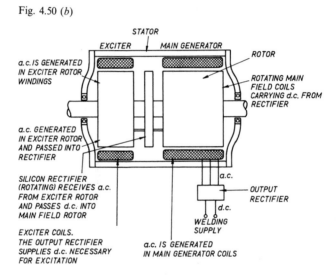

STATOR

EXCITER | MAIN GENERATOR

ROTOR

a.c. IS GENERATED IN EXCITER ROTOR WINDINGS

ROTATING MAIN FIELD COILS CARRYING d.c. FROM RECTIFIER

a.c. GENERATED IN EXCITER ROTOR AND PASSED INTO RECTIFIER

SILICON RECTIFIER (ROTATING) RECEIVES a.c. FROM EXCITER ROTOR AND PASSES d.c. INTO MAIN FIELD ROTOR

a.c.

OUTPUT RECTIFIER

d.c.

WELDING SUPPLY

EXCITER COILS. THE OUTPUT RECTIFIER SUPPLIES d.c. NECESSARY FOR EXCITATION

a.c. IS GENERATED IN MAIN GENERATOR COILS

Fig. 4.51. (*a*) Welding generator with separate excitation. Regulation of controls (1) and (2) gives dual continuous control of voltage and current.

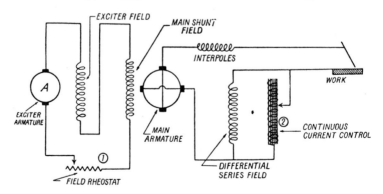

EXCITER FIELD

MAIN SHUNT FIELD

INTERPOLES

WORK

A

EXCITER ARMATURE

MAIN ARMATURE

②

CONTINUOUS CURRENT CONTROL

①

FIELD RHEOSTAT

DIFFERENTIAL SERIES FIELD

run up to speed and their voltages adjusted by means of the shunt field regulators to be equal, they can be connected in parallel by connecting $+$ve terminal to $+$ve, and $-$ve terminal to $-$ve, and the supply taken from the now common $+$ve and common $-$ve terminals. When a load is applied it can be apportioned between either machine by adjustment of the voltage. As the adjustment is made, for example, to increase the voltage of one machine this machine will take an increased share of the load and vice versa (Fig. 4.51*b*).

Fig. 4.51. (*b*) Shunt wound generators connected in parallel. (*c*) Compound wound generators connected in parallel.

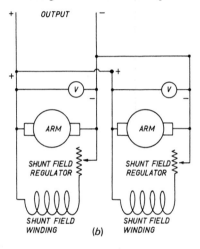

(*b*)

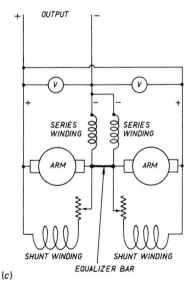

(*c*)

Welding generators, however, are more often compound wound and if connected in parallel as for shunt machines they would not work satisfactorily, because circulating currents caused by any slight difference in voltage between the machines could cause reversal of one of the series fields, and this would lead eventually to one machine only carrying the load. If, however, an equalizer bar is connected from the end of the series field next to the brush connexion on each machine (Fig. 4.51c) the voltage across the series windings of each machine is stabilized and the machines will work satisfactorily. The equalizer connexion should be made when the machines are paralleled as for shunt machines, +ve to +ve and −ve to −ve, and load shared by operation of the shunt field regulators.

Static characteristics of welding power sources

Volt-ampere curves. Variation of the open-circuit voltage greatly affects the characteristics of the arc.

To obtain the volt-ampere curves of a power source:
(1) Set the voltage control to any value.
(2) With the arc circuit open, read the open-circuit voltage on the voltmeter.
(3) Short-circuit the arc.
(4) Vary the current from the lowest to highest value with the current control and, for each value of current, read the voltage. (Voltage will decrease as current increases.) Fig. 4.52a.

Plot a curve of these readings with voltage and current as axes. This curve is a volt-ampere curve and has a drooping characteristic. Any number of curves may be obtained by taking another value of open-circuit voltage and repeating the experiment. The curves in Fig. 4.52b are the results of typical experiments on a small welding generator.

Suppose Fig. 4.52c is a typical curve. When welding, the arc length is continually undergoing slight changes in length, since it is impossible for a welder to keep the arc length absolutely constant. This change in length results in a change in voltage drop across the arc; the shorter the arc the less the voltage drop. The volt-ampere curves shows us what effect this change

Fig. 4.52 (a)

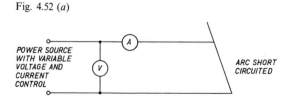

Fig. 4.52 (*b*)

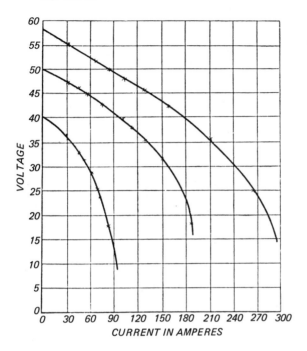

Fig. 4.52 (*c*)

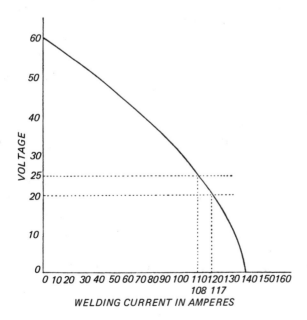

of voltage drop across the arc will have on the current flowing. Suppose the arc is shortened and the drop changes from 25 to 20 V. From the curve, we see that the current now increases from 108 to 117 A.

The steeper the curve is where it cuts the arc voltage value, the less variation in current there will be, and, therefore, there will be no current 'surges' and the arc will be steady and the deposit even. Because the slope of the curve controls the variation of voltage with current this is known as 'slope control'. The dynamic characteristic of a power source indicates how quickly the current will rise when the source is short-circuited.

Variation of current and voltage control. Suppose a current of 100 A is suitable for a given welding operation. If the current control is now reduced, the current will fall below 100 A, but it can be brought back to 100 A by increasing the voltage control. The current control may be again reduced and the voltage raised again, bringing the current again to the same value. At each increase of voltage the volts drop across the arc is increased, so we obtain a different arc characteristic, yet with the same current.

This effect of control should be thoroughly grasped by the operator, since by variation of these controls the best arc conditions for any particular work are obtained.

The curves just considered are known as static characteristics. Now let us consider the characteristics of the set under working conditions; these are known as the dynamic characteristics, and they are best observed by means of a cathode-ray oscilloscope. By means of this instrument the instantaneous values of the current and voltage under any desired conditions can be obtained as a wave trace or graph, called an oscillograph.

The curves drawn in Fig. 4.53 are taken from an oscillograph of the current and voltage variation on a welding power source when the external circuit was being short-circuited (as when the arc was struck) and then open-circuited again.

It will be noticed that the short-circuit surge of current (125 A) is about $1\frac{1}{2}$ times that of the normal short-circuit current (80 A). This prevents the electrode sticking to the work by an excessive flow of current when the arc is first struck, yet sufficient current flows initially to make striking easy. In addition, when the circuit is opened, the voltage rises to a maximum and then falls to a 'reserve' voltage value of 45 V and immediately begins to rise to normal. This reserve voltage ensures stability of the arc after the short-circuit which has occurred and makes welding easier, since short-circuits are taking place continually in the arc circuit as the molten drops of metal bridge the gap.

Motive power for welding generators

Welding generators may be motor- or engine-driven. Sets in semi-permanent positions, such as in workshops, are usually driven by direct current or alternating current motors, and these provide an excellent constant speed drive, since the speed is almost independent of the load. The motor and generator may be built into the same yolk, or may be separate machines. The first method is mostly used in modern machines, as much space is thereby saved.

Fig. 4.53

(a) Curve showing variation in voltage as machine is short-circuited and open-circuited.

(b) Curve showing variation of current due to above variation of voltage.

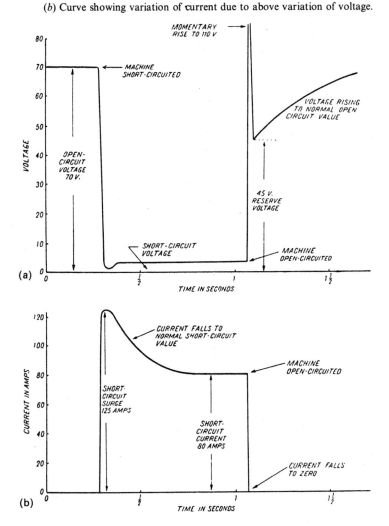

Motors of either type should be fitted with no-volt and overload tripping gear. The former automatically switches off the supply to the motor in the event of a failure of the supply and thus prevents the motor being started in the 'full on' position when the supply is resumed, while the latter protects the motor against excessive overloading, which might cause damage. This operates by switching the motor off when the current taken by the motor exceeds a certain value, which can be set according to the size of the motor.

Main switch and fuses usually complete the equipment of the motor. The motor-driven set may be mounted on wheels or on a solid bed, depending on whether it is required to be portable or not, and the equipment should be well earthed to prevent shock.

Portable sets for outdoor use are usually engine-driven, and this type of set is extremely useful, since it can be operated independently of any source of electric power. The engines may be of the petrol of diesel type and are usually the four-cylinder, heavy duty type with an adequate system of water cooling and a large fan.

A good reliable governor that will regulate the speed to very close limits is an essential feature of the engine. Many modern sets now have an idling device which cuts down the speed of the machine to a tick-over when the arc is broken for a period (which can be adjusted by the operator), sufficient for him to change electrodes and deslag. This results in a considerable saving in fuel and wear and tear and greatly increases the efficiency of the plant.

Direct drive is mostly favoured for welding generators. Belt drive is not very satisfactory, owing to the rapid application of the load when striking the arc causing slip and putting a great strain on the belt, especially at the fastener. V -belt drive sets, however, are used in certain circumstances.

Alternating current welding

Steel fabrication by manual metal arc welding using covered electrodes is now mainly performed using a.c. power sources and this method has certain advantages over the use of d.c. The chief of these are:

(1) The welding transformer (dealt with later) and its controller are very much cheaper than the d.c. set of the same capacity.

(2) There are no rotating parts, and thus no wear and tear and maintenance of plant.

(3) Troublesome magnetic fields causing arc blow are almost eliminated.

(4) The efficiency is slightly greater than for the d.c. welding set.

The following points should be noted concerning a.c. welding:

(1) Covered electrodes must be used. The a.c. arc cannot be used satisfactorily for bare wire or lightly coated rods as with the d.c. arc.

(2) A higher voltage is used than with d.c., consequently the risk of shock is much greater and in some cases, as for example in damp places or when the operator becomes hot and perspires, as in boiler work, a.c. welding can become definitely dangerous, unless care is taken.

(3) Welding of cast iron, bronze and aluminium cannot be done anything like as successfully as with d.c.

The transformer

The supply for arc welding with alternating current is usually from 80 to 100 V, and this may be obtained directly from the supply mains by means of a transformer, which is an instrument that transforms or changes the voltage from that of the mains supply to a voltage of 80 to 100 V suitable for welding. Since a transformer has no moving parts, it is termed a 'static' plant.

The action of the transformer can be understood most easily from the following simple experiment, first performed by Faraday.

An iron ring or core (Fig. 4.54) is wrapped with two *insulated* coils of wire: *A* (called the primary winding) is connected to a source of alternating current, while *B* (called the secondary winding) is connected to a milliammeter with a centre zero, which will indicate the direction of flow of the current in the circuit. With each revolution of the coil of the a.c. generator, the current flows in the primary first from *X* to *Y* and then from *Y* to *X*, and a magnetic flux is set up in the iron core which rises and falls very much in the same way as the hair spring of a watch. This rising and falling magnetic flux, producing a change of magnetic flux in the circuit,

Fig. 4.54

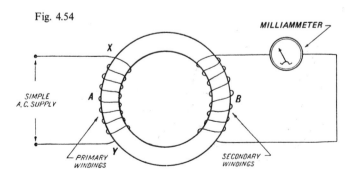

generates in the secondary coil an alternating current, the current flowing in one direction through the milliammeter when the current in the primary is from X to Y and then in the opposite direction when the current in the primary flows from Y to X. There is no electrical connexion between the two coils, and a current generated in the secondary coil in this way, by a current in the primary, is said to be *induced*. Note that we again have the three factors necessary for generation as stated on p. 185: a conductor, a magnetic flux and motion. In this case, however, it is the change in magnetic flux which takes the place of the motion of the conductor, since this latter is now stationary.

Now let us wind a similar ring (Fig. 4.55) with 400 turns on the primary and 100 turns on the secondary, connect the primary to an alternating supply of 100 V, and connect a voltmeter across each circuit.

It will be found that the voltage across the secondary coil is now 25 V.

$$\text{Ratio of } \frac{\text{primary turns}}{\text{secondary turns}} = \frac{4}{1},$$

$$\text{ratio of } \frac{\text{primary voltage}}{\text{secondary voltage}} = \frac{4}{1}.$$

Thus we see that the voltage has been changed in the ratio of the number of turns, or

$$\frac{\text{primary turns}}{\text{secondary turns}} = \frac{\text{primary volts}}{\text{secondary volts}}.$$

This is a simple transformer, and since it operates off one pair of a.c. supply conductors, it is called a *single-phase* transformer. The voltage supplied *to* the transformer is termed the input voltage, while that supplied *by* the transformer is termed the output voltage. If the output voltage is greater than the input voltage, it is termed a *step-up* transformer; while if

Fig. 4.55

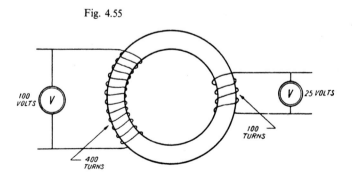

the output voltage is less than the input, it is a *step-down* transformer. Transformers for welding purposes are always step-down, the output voltage being about 85 V. Single-operator transformers have two output tappings of 80 and 100 V, the higher voltage being suitable for light gauge sheet welding (Fig. 4.56). The input voltage to transformers is usually 415 or 240 V, these being the normal mains supply voltages. The alternating magnetic field due to the alternating current in the windings would generate currents in the iron core if it were solid. These currents, known as eddy currents (or Foucault currents), would rapidly heat up the core and overheat the transformer. To prevent this, the core is made up of soft iron laminations varnished with insulating varnish so as not to make electrical contact with each other, and clamped tightly together with bolts passing through insulating bushings to prevent the bolts carrying eddy currents. In this way losses are reduced and temperatures kept lower. Eddy currents are also used for induction heating in the electric induction furnace.

Since the power output cannot be greater than the input (actually it is always less because of losses in the transformer), it is evident that the current will be transformed in the opposite ratio to the voltage. For example, if the supply is 400 V and 50 A are flowing, then if the secondary output is 100 V, the current will be 200 A (Fig. 4.57).

Actually, the output current would be slightly lower than this, since the above assumes a 100% efficient transformer. A transformer on full load has an efficiency of about 97%, so the above may be taken as approximately true.

The highly magnetizable silicon iron core of the transformer is made up

Fig. 4.56

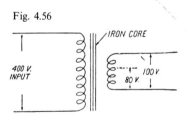

Fig. 4.57

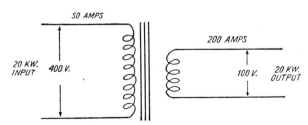

of laminations bolted together and the coils fit over the these (Fig. 4.58). It will be observed that the magnetic circuit is 'closed', that is, the flux does not have to traverse any air gap.

The single-operator welding transformer is made on this principle and is available in sizes up to 450–500 A.

The transformer may be of the dry type (air cooled) or it may be immersed in oil, contained in the outer container. Oil-immersed transformers have a lower permissible temperature rise than the dry type and, therefore, their overload capacity (the extent to which they may be used to supply welding currents in excess of those for which they are rated) is much smaller.

Excessive variation in the supply voltage to a transformer welding set may affect the welding operation by causing a variation in welding current and voltage drop across the arc and the open circuit voltage (OCV). For 80 V the ratio is 80/410. If the supply falls to 380 V the OCV of the secondary will now be 80/410 × 380 or approximately 76 V. This fall will reduce arc current by about 4 A if the original current was 100 A, and the arc voltage by 1–1.5 V, which will have some effect on welding conditions. Many modern welding units have thyristor regulators by which the arc voltage is kept constant irrespective of any variations of mains voltage, keeping welding conditions constant. Voltage control is by transductor, which gives infinite adjustment.

By law the supply authorities must not vary the supply voltage from that specified by more than 6% so that variation in the case of a specified 410 V supply is from 385–434 V, but this may be exceeded under adverse conditions. Small voltage variations have little effect on welding conditions.

Current control

Current control may be by tapped reactor (choke), flux leakage reactor, saturable reactor, leakage reactance moving coil, or thyristor

Fig. 4.58

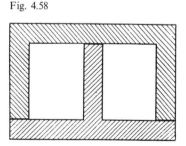

controlled transformer. In the latter the circuit is controlled by one knob
and is stepless (see Thyristor).

Inductive reactor or choke

An inductive reactor or choke consists, in its simplest form, of a
coil of insulated wire wound on a closed laminated iron core (Fig. 4.59).
When an e.m.f. is applied to the coil and the current begins to flow, the iron
core is magnetized. In establishing itself, this flux cuts the coil which is
wound on the core and generates in it an e.m.f. in the opposite direction to
the applied e.m.f. and known as the 'back e.m.f.'. It is the result of magnetic
induction, and its effect is to slow down the rate of rise of current in the
circuit so that it does not rise to its maximum value as given by Ohm's law
immediately the e.m.f. is applied; in a very inductive circuit the rise to
maximum value may occupy several seconds.

The inductance of the circuit is proportional to the square of the number
of turns of wire on the coil, so that increasing the number of turns greatly
increases the inductive effect. When the current is fully established, there is
energy stored in the magnetic circuit by virtue of the magnetic flux in the
iron core. When the circuit is broken, the lines of force collapse, and in
collapsing cut the coil and generate an e.m.f. in the opposite direction to
that when the circuit was made, and which now tends to maintain the
current in the original direction of flow. The energy of the magnetic flux is
thus dissipated and may cause a spark to occur across the contacts where
the circuit is being broken.

The direction of the induced e.m.f. in an inductive circuit is given by
Lenz's law, which states: 'The direction of the induced effect in an inductive
circuit always opposes the motion producing it.'

If an alternating current is flowing in the coil, the current will reverse
before it has time to reach its maximum value in any given direction, and

Fig. 4.59. Tapped choke.

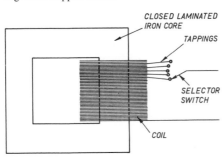

the more inductive the circuit the lower will the value of the current be, so that the effect is to 'choke' the alternating current. If the coil has tappings taken to a selector switch, the inductance of the circuit and hence the amount by which it can control or 'choke' the current can be varied (Fig. 4.59). Another method of varying the inductive effect is to vary the iron circuit so that the flux has a 'leakage path' other than that on which the coils are wound (Fig. 4.60), thus varying the strength of the magnetic flux in the core and hence the inductive effect.

The tapped reactor or choke is used to control the current in metal arc welding a.c. welding units. It can only be used on a.c. supplies and does not generate heat as does a resistor used for control of direct current. Any heat generated is partly due to the iron core, and partly due to the $I^2 R$ loss in the windings (Fig. 4.61).

Leakage reactance, moving coil current regulation

Stepless control of the current is achieved in this method by varying the separation of the primary and secondary coils of the transformer. The coils fit onto the iron circuit as shown in Fig. 4.62. The secondary coil supplying the welding current is fixed and the primary coil can be moved up and down by means of a screw thread and nut, operated by a winding handle mounted on top of the unit. As the primary coil is wound so as to approach the secondary coil the inductive reactance is reduced and the current is increased and vice versa, so that the highest current values are when the coils are in the closest proximity to each other. The moving coil carries a pointer which moves over two scales, one high current values and one low values, the different scales being obtained by two tappings on the secondary coil. The inductive reactance does not vary

Fig. 4.60. Flux leakage control.

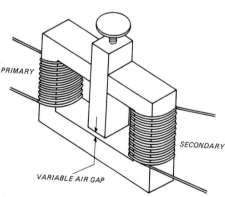

PRIMARY

SECONDARY

VARIABLE AIR GAP

directly with the separation of the coils, and current values get rapidly greater as the coils get near to each other.

The merit of this method of current control is that there is no other item of equipment other than the transformer, thus reducing the initial cost, and there is stepless control operating with the simplest of mechanisms.

Magnetic induction and saturation

If a coil of insulated wire is wound on an iron (ferromagnetic) core and a current is passed through the coil, a magnetic flux is set up in the iron core. The strength of this flux depends upon the current in amperes and the number of turns of wire on the coil, that is, upon the ampere-turns (AT) so that the magnetizing force (H) is proportional to $A \times T$. If a graph is drawn between the magnetizing force and the flux density (B) in the core it is known as B/H or magnetization curve (Fig. 4.63). It will be noticed that the flux density rises rapidly to X with small increases of H and then begins to flatten out until at Y further increases of H produce no further increase of flux density B. At the point Y the core is said to be magnetically saturated.

Use is made of this to control the current in power units, being known as the saturable reactor method.

Fig. 4.61

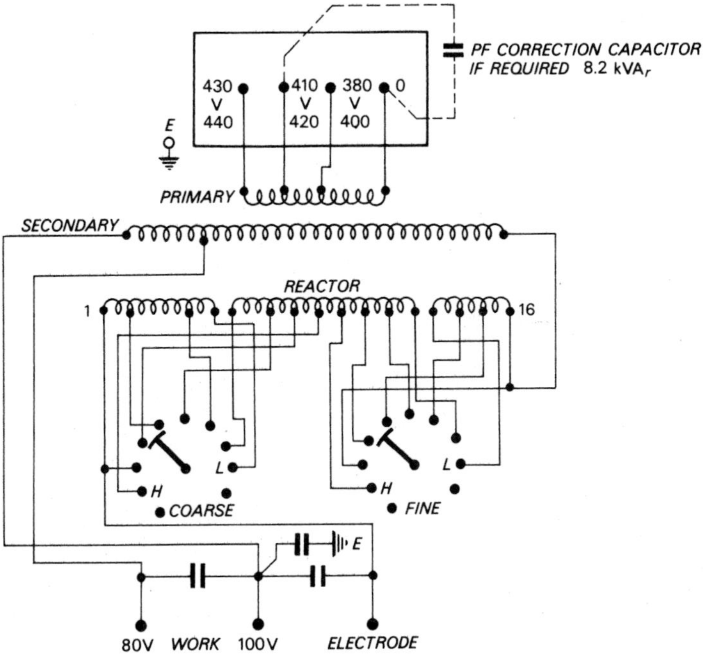

A coil A is wound on one limb of a closed laminated iron core and carries d.c. from a bridge connected rectifier X and controlled by a variable resistor Y (Fig. 4.64). A coil carrying the main welding current is wound on the other limb. When there is no current through A, the coil B will have maximum reactance because there is no flux in the core due to A and the welding current will be a minimum. As the current in A is increased by the control Y, magnetic saturation can be reached, at which point reactance is a minimum and the welding current will be a maximum represented by maximum voltage on V so that between these limits, accurate stepless control of the welding current is achieved.

Behaviour of a capacitor in an a.c. circuit

When a capacitor is connected to a d.c. source a current flows to

Fig. 4.62

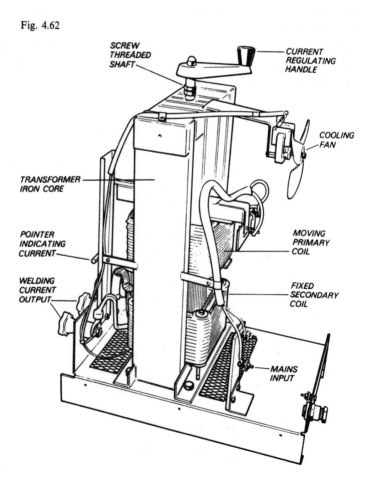

charge it and no further current flows, the capacitor preventing or blocking the further flow of current.

If the capacitor is now connected to an a.c. source the change of polarity every half-cycle will produce the same change of polarity in the capacitor so that there is a flow of current, first making one plate positive and then half a cycle later making it negative, and the current which is flowing in the circuit, but not through the capacitor, is equal to the charge current. Hence a capacitor behaves as an infinitely high resistor in a d.c. circuit preventing flow of current after the initial charge, while in an a.c. circuit the current flows from plate to plate, not through the capacitor but around the remaining part of the circuit (Fig. 4.65).

Fig. 4.63

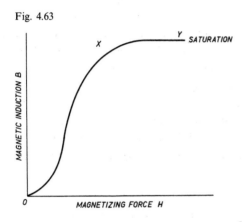

Fig. 4.64. Experiment with saturable reactor control.

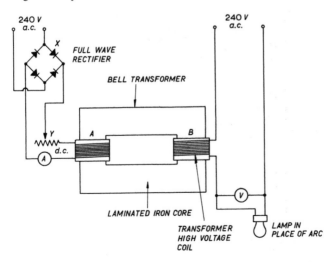

Phase of current and voltage in an a.c. circuit

When a voltage is applied to a circuit and a current flows, if there is an inductive effect in the circuit the current will fall out of step with the voltage and lag behind it, rising and falling at the same frequency, but lagging a number of degrees behind. If there is capacitance in the circuit, the current will lead the voltage. Zero values of voltage and current do not occur together (Fig. 4.66*b*) and there is always some energy available, so that in a welding circuit the arc is easier to strike and maintain when a tapped reactor, for example, used for current control is in the circuit, since this produces an inductive effect.

Inductive reactance

In any circuit, the effect of inductance is to increase the apparent resistance of the circuit. This effect is termed inductive reactance and if there is capacitance in the circuit the effect is known as capacitive reactance.

The unit of inductance is the henry (H). A circuit has an inductance of 1 henry if a current, varying at the rate of 1 ampere per second, induces an e.m.f. of 1 volt in the circuit.

If an alternating e.m.f. of V volts at frequency f Hz is applied to a circuit of inductance L henrys and a current of I amperes flows, the volts drop V across the inductor $= 2\pi LfI = IX_L$, where $X_L = 2\pi fL$. Comparing this with the volts drop V across a resistor R ohms carrying a current of I

Fig. 4.65. Flow of a.c. in a circuit containing a capacitor.

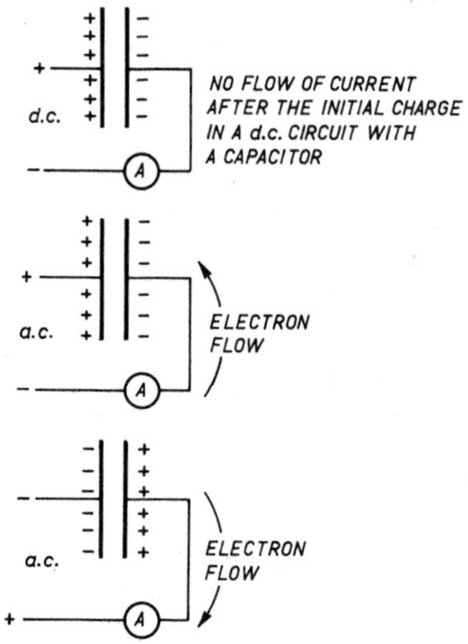

amperes, $V = I \times R$ so that X_L takes the place of R and is known as the inductive reactance.

Impedance

If a circuit contains resistance and inductance in series (Fig. 4.67a), the current I amperes flows through both inductance and resistance and there will be a volts drop IR across the resistor, in phase with the voltage, and a volts drop across the inductor, 90 ° out of phase with the voltage. This

Fig. 4.66. (a) Voltage and current in phase. Both pass through zero and maximum values at the same time. (b) Voltage and current out of phase due to inductance in circuit current lagging 45 ° behind voltage. Current and voltage now do not pass through zero and maximum values together.

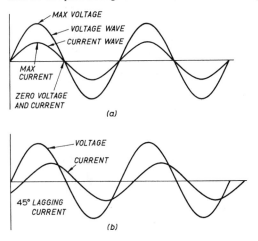

Fig. 4.67. Circuit with resistance and inductance in series.

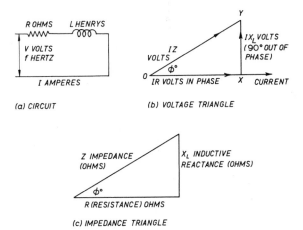

can be represented by a phasor* diagram (Fig. 67*b*) using the current common to resistance and inductance as a reference phasor.

OX represents the volts drop IR in phase with the current.

XY represents the volts drop IX_L across the inductor, 90° out of phase, leading the current.

Then OY is the resultant voltage across the circuit, leading the current by an angle $\varphi°$ so that the current *lags* behind the voltage $\varphi°$, and if $OY = I \times Z$ where Z is the resultant effect of R and X_L, Z is termed the impedance. If the sides of the voltage triangle are all divided by I we have (Fig. 4.67*c*) the impedance triangle, and from this $Z^2 = R^2 + X_L^2$, or

$$\text{impedance}^2 = \text{resistance}^2 + \text{inductive reactance}^2.$$

The unit of capacitance is the farad (F), see p. 176. If an alternating e.m.f. of V volts at frequency f Hz is applied to a circuit of capacitance C farads and a current of I ampere flows:

$$V = \frac{I}{2\pi f C} = IX_C, \text{ where } X_C = \frac{1}{2\pi f C}$$

and is termed the capacitive reactance of the circuit.

Let a current of I amperes flow in a circuit containing a resistor of R ohms and a capacitor of capacitance C farads when an e.m.f. of V volts is applied at frequency f Hz (Fig. 4.68*a*). The voltage across the resistor, in phase with the current, is IR volts, whilst the volts drop across the capacitor is IX_C volts, the voltage lagging the current by 90°. The phasor diagram in Fig. 4.68*b* represents this with OX as the reference current phasor.

OX represents the volts drop IR, in phase with the current, XY represents the volts drop across the capacitor, lagging the current by 90°. Then OY represents the resultant voltage across the circuit, lagging the current by an angle $\varphi°$ and $OY = IZ$, where Z is the impedance of the circuit. Fig. 4.68*c* is the impedance triangle where $Z^2 = R^2 + X_C^2$. Impedance is measured in ohms (apparent).

---

* A scalar is a quantity which has magnitude only, e.g. mass or temperature. A vector is a quantity which has magnitude and direction, e.g. velocity or force. Phasors are rotating vectors. They are easier to draw than the more complicated wave form diagrams and are added or subtracted as are vectors.

  In the phasor diagram the phasors are drawn to scale to represent the magnitude of the quantity (e.g. volts or amperes) and the angle between the phasors represents the phase displacement, the whole rotating counter-clockwise at an angular velocity measured in radians per second. Spokes on a bicycle wheel provide an analogy. Irrespective of the angular velocity of the wheel the angle between any two given spokes (representing the phasors) remains the same.

Resistance, inductance and capacitance in series

In a circuit such as that used for TIG welding there is inductance in the current controls, capacitance used for blocking the d.c. component, and the resistance of the circuit in series (Fig. 4.69a). The phasor diagram (Fig. 4.69b) shows that IX_L and IX_C are in opposite directions (anti-phase) and the resultant reactance is $X_L - X_C$, since inductance is usually larger than capacitance. It will be noticed that the impedance is always greater than the ohmic resistance so that, if d.c. is applied to a circuit designed for a.c., excess currents will flow. Fig. 4.69c is the voltage triangle and Fig. 4.69d the impedance triangle, and from this $Z^2 = R^2 + (X_L - X_C)^2$, or

$$\text{impedance} = \sqrt{[(\text{resistance})^2 + (\text{resultant reactance})^2]} \text{ ohms.}$$

Fig. 4.68. Circuit with resistance and capacitance in series.

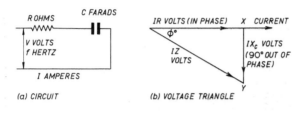

Fig. 4.69. Circuit with resistance, inductance and capacitance in series.

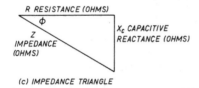

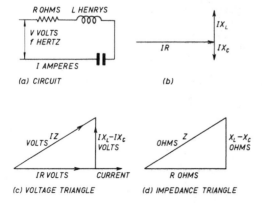

Power factor

Suppose that a current in a circuit is lagging by an angle $\varphi°$ behind the voltage. To find the true power in the circuit the phasor diagram (Fig. 4.70*a*) is drawn, in which OV represents the voltage and OB the current lagging the voltage by an angle $\varphi°$. The length of OB is drawn to represent the current I in amperes and this is resolved into two components, AO in phase with the voltage and AB 90° out of phase with the voltage.

In the triangle AOB, $AO/OB = \cos\varphi$, therefore $AO = OB \cos\varphi$ and $AB = OC$ and $AB/OB = \sin\varphi$, therefore $AB = OB\sin\varphi$

The component in phase with the voltage which is the power component is $I\cos\varphi$ amperes, while the component 90° out of phase (the reactive component) which is wattless is $I\sin\varphi$ and produces no useful power. Thus the power in a circuit which is the product of voltage and current is $VI\cos\varphi$. The cosine of any angle cannot be greater than 1, so that the power in a reactive circuit is always less than the product of the volts and amperes. If $\varphi = 0°$, that is, the current and voltage are in phase, the power is $VI\cos 0° = VI$, since $\cos 0° = 1$. If $\varphi = 90°$, the power is $VI\cos 90°$, and since $\cos 90° = 0$, the power is zero so that the reactive component produces no power in the circuit.

The factor $\cos\varphi$ is known as the *power factor*. The more inductive the circuit, the more will the current be out of phase with the voltage and the greater will be the angle φ so that $\cos\varphi$ gets smaller and the power becomes less.

If an a.c. welding supply is, for example, 40 V with a current of 120 A at a power factor of 0.7 lagging, the power in the circuit is

$$VI\cos\varphi = 40 \times 120 \times 0.7 = 3360 \text{ W} = 3.36 \text{ kW}.$$

If the power factor were unity or the circuit d.c. the power would be

$$40 \times 120 \times 1 = 4800 \text{ W} = 4.8 \text{ kW}.$$

Thus in any a.c. circuit, the product of the volts and amperes gives the apparent power in volt-amperes or kilovolt-amperes (kVA), while the true power in kilowatts is kVA × power factor $\cos\varphi$.

Fig. 4.70

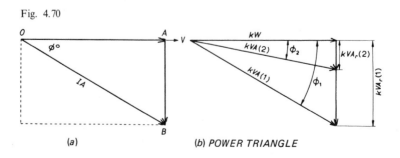

(a)   (b) POWER TRIANGLE

If a power triangle is drawn, the kW are in phase with the voltage and the kVA are in phase with the current, and kVA_r represents the reactive component of the power.

In Fig. 4.70b the power triangle is drawn for two different angles of phase difference, φ_1 and φ_2, and the reactive components are kVA_r (1) and kVA_r (2) respectively, so that by reducing the angle of lag from φ_1 to φ_2, the reactive kVA is reduced from kVA_r (1) to kVA_r (2) and the total kVA is reduced from kVA (1) to kVA (2) whilst the true power is represented in each case by kW.

Since the kVA is proportional to the current flowing, the supply current can be reduced for a given kW by reducing the angle of lag φ, thus reducing the power (I^2R) loss in the supply cables and transformers.

To encourage consumers to have as high a power factor as possible by, for example, the installation of banks of capacitors, supply authorities have a tariff based on kVA_r maximum demand which must operate for a given period before registering on a dial and upon which maximum demand the tariff is based, so that a consumer with a low power factor and thus a high kVA_r pays more per unit for energy. Welding equipment, because of the transformers and chokes, tends to give a low power factor.

Fitting of power factor improvement capacitors is an important consideration especially when there is a large transformer load as in the welding industry. Capacitors for power factor improvement are rated in kVA_r, the r indicating the reactive (out of phase) component of the power. For example, suppose a typical transformer has a maximum output welding current of 450 A with an input of 440 V, 90 A and a lagging power factor of 0.46 (current lagging behind the voltage by 62°). If an 8.2 kVA_r bank of capacitors is installed, the input current will now fall to 75 A at 440 V, improving the power factor to 0.57 (lagging 55°). In terms of power, the original input was approximately 40 kVA, but with the capacitors improving the power factor the new power input has been reduced to 33 kVA, giving a considerable reduction in power consumed. Thus the saving in energy cost would soon pay for the capital outlay of the power factor improvement capacitors (Fig. 4.71).

Three-phase welding supply

For convenience in transmission and distribution, alternating current is supplied on the 'three-phase' system. The alternators have three sets of coils set at an angle of 120° to each other, instead of only one coil as on the simple alternator which we considered. These coils can be connected, as shown in Fig. 4.72, and the centre point, termed the star point, is where the beginning of each coil is connected, and the wire from

Fig. 4.71

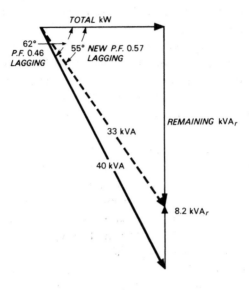

Fig. 4.72. (*a*) Connexions of a simple single-phase single-operator transformer set. (*b*) three-phase, four-wire system.

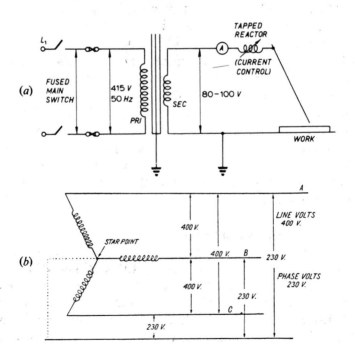

this point is termed the neutral. *A*, *B* and *C* are the lines. The voltage between *A* and *B*, *B* and *C*, *C* and *A* is termed the line voltage and is usually for supply purposes between 440 and 400 V. The voltage between any one of the lines and the neutral wire, termed the phase voltage, is only $1/\sqrt{3}$ of the line voltage, that is, if the line voltage is 400 V the voltage between line and neutral is $400/\sqrt{3} = 230$ V, or if the line voltage is 415 V the phase to neutral voltage is 240 V.

Welding supplies for more than one welder are supplied by multi-operator sets from the above type of mains supply.

Welding transformers

Welding transformers can be single-phase or three-phase. Single-phase transformers are connected either across two lines with input voltage 380–440 V or across one line and neutral when the voltage is 220–250 V. Evidently the single-phase transformer is an unbalanced load since all three lines are not involved. To balance the load equally on the three lines is not possible in welding using three single-phase transformers (Fig. 4.73a) since the welders are seldom all welding together and using the same current, so that in practice balance is never realized. Three-phase transformers on the other hand give a better balancing of the load even when only one welder is operating (Fig. 4.73b). Single-phase transformers are available for single-operator welding with a variety of outputs. As the input voltage is reduced the input current rises for the same power output so that it is usually the smaller output units which are made for line-to-neutral (240 V) connexion. Larger units are connected across two lines to keep the input current down.

The open circuit voltage (OCV) depends upon the particular transformer. Many transformers have 80 to 100 OCV selected as required, for example giving 50–450 A at 80 OCV and 60–375 A at 100 OCV. Other

Fig. 4.73 (a)

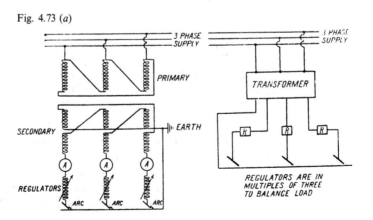

transformers may have a lower OCV of 70 V or even 50 V. The electrode classification indicates the types of electrode coverings suitable for the various OCV, and the striking voltage required is always given with the instructions for use of a particular type of electrode.

Smaller transformers with output currents of about 200 A maximum are air-convection-current cooled: larger units are forced draught (fan assisted) while most of the largest units are oil cooled.

As is the case with most electrical machines the duty cycle is important in order to keep the temperature rise within permissible limits. Although the maximum current specified for a given transformer may be say 200 A, this rating may be only possible for a 25% duty cycle, with, for example, 180 A at 30% and 100 A at 100% duty cycle (continuous welding). When choosing a unit therefore it is important to estimate the average current settings that will be used and the approximate duty cycle, so that a large enough unit can be selected, i.e., one that will perform the work without excessive temperature rise.

Current control can be by tapped choke, leakage reactance moving coil, thyristor, and for fine current settings can have 40–50 steps with two selectors, one coarse, one fine, while the leakage reactance moving coil type can have a continuously variable current control operated by hand wheel or lever.

When transformer units are to be used for TIG welding in conjunction with an HF unit they are fitted with an HF protection circuit because of the high voltage involved. Multi-operator equipment is often used in larger welding establishments at a saving in capital cost, with 6, 8 or 12 welders being supplied from one three-phase transformer, each with his own current regulator. The sizes vary from those for 6 welders each with a maximum welding current of 350 A to the largest units for 12 welders with a maximum current of 450 A each and a rating of 486 kVA (Fig. 4.74).

Fig. 4.73 (*b*)

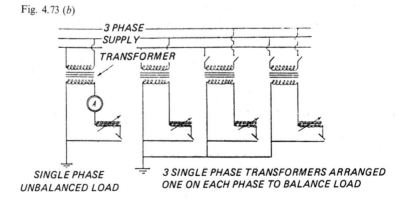

SINGLE PHASE
UNBALANCED LOAD

3 SINGLE PHASE TRANSFORMERS ARRANGED
ONE ON EACH PHASE TO BALANCE LOAD

When using a transformer in which the current is selected by a coarse and fine tapping switch, the current should not be altered whilst the welding current is flowing since the arcing which occurs as the selector passes from stud to stud damages the smooth surface of the contact stud. If the transformer is oil cooled, the quenching action of the oil prevents serious arcing, but some oil may be carbonized and this will eventually cause a deterioration of its insulating qualities.

In addition, the three-phase transformer is cheaper to manufacture and install than three single-phase transformers and, because of this, is often found wherever many welders have to be supplied, as in shipyards and engineering works. Each welder has his own current regulator, as in the single-phase set, but it is not mobile as is the single-operator set. Details of the electrical equipment used in TIG, MIG and CO_2 processes are included under their respective headings.

Note. See also BS 638, *Arc welding plant, equipment and accessories.*

Parallel operation of welding transformers

Welding transformers of similar type can be connected in parallel to give a greater current output than could be provided by either of them used singly. The transformers should have their primaries connected across the same pair of lines and the output welding voltages should be the same in order to prevent circulating currents flowing in the secondary windings

Fig. 4.74. Three-operator set, showing the single three-phase transformer and three welding regulators.

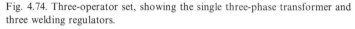

before they are connected to a load. There is no problem of phase rotation but the output should be checked for 'polarity'.

This is done by connecting both 'work' terminals together and placing a voltmeter across the 'electrode' terminals as shown in Fig. 4.75. If the voltmeter reads zero the transformers have similar polarity and the electrode terminals can be connected together, and welding performed from the paralleled units. If the voltmeter reads twice the normal output voltage the polarity is reversed and the connexions to one pair of secondary terminals (work and electrode) should be reversed, when the test should show zero voltage and the transformers can now be paralleled.

Earthing

If a person touches a 'live' or electrified metal conductor, a current will flow from this conductor, through the body to earth, since the conductor is at higher electrical pressure (or potential) than the earth. The shock that will be felt will depend upon how much current passes through the body and this in turn depends upon (1) the voltage of the conductor, (2) the resistance of the human body, (3) the constant resistance between body and earth.

The resistance of the human body varies considerably and may range from 8000 to 100 000 ohms, while the contact resistance between body and earth also has a wide range. Resistance to earth is high if a person is standing on a dry wooden floor and thus a low current would pass through the body if a live conductor is touched, while if a person is standing on a wet concrete floor and touches a live conductor with wet hands the resistance to earth is greatly lowered, a larger current would pass through the body and consequently a greater shock would be felt. It may be stated here that care

Fig. 4.75. Paralleling of welding transformers, polarity test.

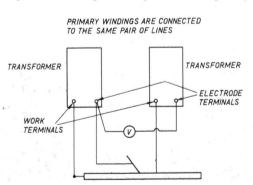

should be taken to avoid shock when welding in damp situations, especially with a.c. The operator can wear gloves and thus avoid touching the welding terminals with bare hands and he can stand on dry boards.

Most electrical apparatus, such as motors, switch gear, cables, etc., is mounted in, or surrounded by, a metal casing, and if this should come into contact, through any cause whatever, with the live conductors inside, it will then become electrified and a source of danger to any one touching it.

To prevent this danger, *all* metal parts of electrical apparatus *must* be 'earthed', that is, must be connected with the general mass of the earth so that at all times there will be an immediate and safe discharge of energy. Good connexion to earth is essential. If the connexion is poor, its resistance is high and a current may follow an easier alternative path to earth through the human body if the live metal part is touched.

For earthing of electrical installations in houses, the copper pipes of the cold water system are sometimes satisfactory since they are sufficient to carry to earth currents likely to be met with in this type of load.

Connexion from the 'earthing system', as it is termed, to earth is made in various ways. Earth plates of cast iron or copper, 1–1.5 m square and buried 1.5–2 m deep, are in general use in this country. They are surrounded by coke and the area around is copiously watered. Tubes, pipes, rods and strips of copper driven deep into the ground are used both in this country and in the USA and the area round them is frequently covered with common salt and again copiously watered.

It is evident that the 'earthing system' must be continuous throughout its length and must connect up and make good contact with every piece of metal likely to come into contact with live conductors. In factories and workshops the cables are carried in steel conduits and this forms the earthing system, the conduit making good contact with all the apparatus which it connects. Any metal part which may become live discharges to earth, through the continuous steel tubing system. To ensure that connexion to earth is well made, extra wires of copper with terminal lugs attached are connected from the conduit to the metal parts of apparatus such as motors and switch gear and ensure a good 'bond' in case of poor connexion developing between the conduit and the metal casing of the apparatus. In the case of portable apparatus such as welding transformers, regulators, welding dynamos (motor driven), drills, hand lamps, etc., an extra earthing wire is run (sometimes included in the flexible tough rubber supply cable) and makes good connexion from the metal parts of the portable apparatus to the main earthing system.

When steel wire or steel tape armoured cable is used, the wire or tape is utilized as the earthing system. In all cases extra wires are run whenever necessary to ensure good continuity with earth, and the whole continuous system is then well connected to the earth plate by copper cables.

In a.c. welding from a transformer it is usual to earth one of the welding supply terminals in addition to the metal parts of the transformer and regulator tanks. This protects the welder in the event of a breakdown in the transformer causing the mains supply pressure to come into contact with the welding supply.

Low voltage safety device

As we have seen, if a welder is working in a damp situation or otherwise making good electrical contact through his clothes or boots with the work being welded (as for example inside a boiler or pressure vessel) and he touches a bare portion of the electrode holder or uses bare hands to place an electrode in the holder, his body is making contact across the open circuit voltage of the supply.

If this is d.c. at about 50–60 OCV practically no effect is felt but if the supply is from a transformer at say 80 OCV, this is the rms value and the peak of this is about 113 V so that the welder will feel an electric shock, its severity depending upon how good a contact is being made between electrode and work by the welders body. In some cases the shock can be severe enough to produce a serious effect.

A safety device is available which is attached to the transformer welding unit and consists of a step-down transformer and rectifier giving 25 V d.c. with contactors and controls.

When the transformer is switched on, a d.c. voltage of 25 V appears across electrode holder and work terminals. When the electrode is struck on the work and the circuit completed a contactor closes and the 80 OCV of the transformer appears between electrode and work and the arc is struck. Immediately the arc is extinguished the 25 V d.c. reappears across electrode and work terminals thus giving complete safety to the welder. Green and red lights indicate low volts and welding in progress respectively.

5

Inspection and testing of welds

During the process of welding, faults of various types may creep in. Some, such as those dealing with the quality and hardness of the weld metal, are subjects for the chemist and research worker, while others may be due to lack of skill and knowledge of the welder. These, of course, can be overcome by correct training (both theoretical and practical) of the operator.

In order that factors such as fatigue may not affect the work of a skilled welder, it is evidently necessary to have means of inspection and testing of welds, so as to indicate the quality, strength and properties of the joint being made.

Visual inspection, both while the weld is in progress and afterwards, will give an excellent idea of the probable strength of the weld, after some experience has been obtained.

Inspection during welding

Metal arc welding. The chief items to be observed are: (1) rate of burning of rod and progress of weld; (2) amount of penetration and fusion; (3) the way the weld metal is flowing (no slag inclusions); (4) sound of the arc, indicating correct current and voltage for the particular work.

Oxy-acetylene welding. The chief items are: (1) correct flame for the work on hand; (2) correct angle of blowpipe and rod, depending on method used; (3) depth of fusion and amount of penetration; (4) rate of progress along the joint.

The above observations are a good indication to anyone with experience what quality of weld is being made, and this method finishes one of the best ways of observing the progress of welders when undergoing training.

Inspection after welding

Examination of a weld on completion will indicate many of the following points:

(1) Has correct fusion been obtained between weld metal and parent metal?

(2) Is there any indentation, denoting undercutting along the line where the weld joins the parent metal (line of fusion)?

(3) Has penetration been obtained right through the joint, indicated by the weld metal appearing through the bottom of the V or U on a single V or U joint?

(4) Has the joint been built up on its upper side (reinforced), or has the weld a concave side on its face, denoting lack of metal and thus weakness?

(5) Does the metal look of a close texture or full of pinholes and burnt?

(6) Has spatter occurred, indicating too high a current or too high a voltage across the arc or too long an arc?

(7) Are the dimensions of the weld correct, tested, for example, by gauges such as shown in Fig. 5.1?

A study of the above will indicate to an experienced welder what faults, if any, exist in the work and then provide a rapid and useful method of ensuring that the right technique of welding is being followed.

A very useful multi-purpose pocket-size welding gauge has been designed by the Welding Institute. It is of stainless steel and enables the following measurements to be taken in either metric or Imperial units: material thickness up to 20 mm; prepation angle 0–60°; excess weld metal capping size; depth of undercut and of pitting; electrode diameter; fillet weld throat size and leg length and high–low misalignment.

Visual inspection, however, has several drawbacks. Take, for example, the double V joint shown in Fig. 5.2. It will obviously be impossible to observe by visual means whether penetration has occurred at the bottom of the V except at the two ends.

Fig. 5.1. Weld test gauges.

 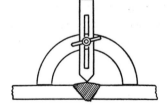

A great variety of methods of testing welds are now available and, for convenience, we can divide them into two classes: (1) non-destructive, (2) destructive.

Destructive tests are usually carried out either on test specimens made specially for the purpose, or may even be made on one specimen taken as representative of several similar ones.

Destructive tests are of greatest value in determining the ultimate strength of a weld and afford a check on the quality of weld metal and skill of the operator. (Visual inspection obviously falls under the heading of non-destructive tests.)

Non-destructive tests (NDT)

(1) Penetrate fluid and visual inspection.
(2) Magnetic (*a*) magnetic particle, (*b*) search coil.
(3) X-ray.
(4) Gamma-ray.
(5) Ultrasonic.
(6) Application of load.

Penetrant fluid. The surface is cleaned and the dye penetrant fluid is painted or sprayed on the area to be examined. The fluid is allowed to penetrate into any defects such as cracks and crevices and the surplus is removed. A developer powder is sprayed on to the surface and soaks up the penetrant leaving a stain indicating the defect. The surface can also be viewed under ultraviolet light in darkened conditions, when the fluorescent penetrant glows, indicating the crack.

Surface scratches may mask the result and penetrant contamination of the crack occurs but the method is used as an addition to X-ray or gamma-ray inspection.

Magnetic tests for magnetizable specimens. Surface defects only.
(*a*) The specimen under test is magnetized using a low voltage transformer and two probes for making contact with the speci-

Fig. 5.2

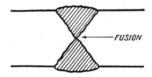

men, and to enable the flux to be varied in the specimen. Iron filings in a finely divided or colloidal state are applied as an ink or as a powder and the flux is distorted at the crack or other fault with magnetic poles being formed. Probe positions will give a flux either with the weld or across it as shown in Fig. 5.3. Examination can also be performed with ultraviolet light and fluorescent ink.

(*b*) The specimen is magnetized as before and search coils, connected to a galvanometer which measures small currents, are moved over the specimen. If a crack exists in the specimen, the change of magnetic flux across it will cause a change in the current in the search coil and is indicated by fluctuations in the galvanometer needle. This method has advantages over method (*a*) in that the surface need not be machined and that defects just below the surface are indicated.

X-rays and gamma rays

X-rays are an electromagnetic radiation delivered in 'quanta' or parcels of energy as opposed to continuous delivery. They move at the speed of light in straight lines; are invisible; are not deviated by a lens; ionize or liberate electrons from matter through which they can pass and they destroy living cells. They are generated by an X-ray tube, described later.

Gamma-rays are similar to X-rays but differ in wavelength, X-rays having a continuous or broad spectrum while gamma-rays are made up of

Fig. 5.3. Position of probes showing magnetic flux.

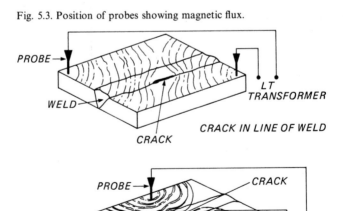

isolated wavelengths and have a line spectrum depending upon the element used. Iridium, for example, has two distinct types of atoms, one with a mass number of 191 and the other with a mass number of 193. The latter has two extra neutrons in its nucleus. These are written 191 Ir and 193 Ir and are isotopes of iridium. If the stable isotope is bombarded with neutrons in a nuclear reactor (such as at Harwell) an additional neutron is induced into the nucleus and the isotope becomes unstable and is termed a radioactive isotope or radioisotope. These unstable isotopes suffer radioactive decay or change into the stable form over a period of time and the type of radiation (wavelength) and the period of time for which it is given out determines its suitability for a particular use. Some isotopes emit radiation at a single level; for example cobalt-60 emits at two energy levels near each other. Others emit a broader spectrum comparable with that from an X-ray tube. Iridium-192 has 16 differing energy levels and gives better contrast with specimens of varying thickness comparable with that of an X-ray distribution.

Fig. 5.4 shows the wavelengths of e.m. radiations (spectra) given in

Fig. 5.4. Electromagnetic spectrum showing position of X-rays and gamma-rays.

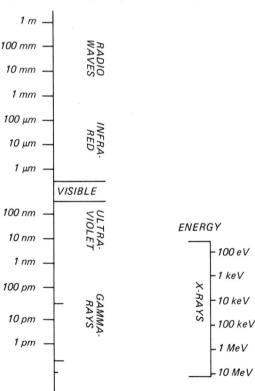

metres (m); micrometres, ($1\mu m = \frac{1}{1000}$ mm); nanometres (1 nm $= \frac{1}{1000}$ μm); and in ångström units (1 Å $= \frac{1}{10}$ nm).

Radioactive decay, in which the isotope emits radiation to attain the stable state, may vary from a fraction of a second to hundreds of years. The decay rate cannot be speeded up nor slowed down and obeys an exponential law, being proportional to the number of radioactive nucleii so that complete decay never occurs as some radioactive nucleii are always left. For this reason decay of a radioactive isotope is expressed in terms of its half life, which is the period of time for the number of nucleii to decay to half that number. Those chosen for radiographic testing of welded joints may vary from a few weeks (e.g. thulium-170) to some years (e.g. cobalt-60). The table gives the radioactive isotopes in general use for testing welds.

The unit of radioactivity is the curie (Ci), also millicurie (mCi) and microcurie (μCi). This is the amount of radioactive material in which 3.7×10^{10} disintegrations take place per second.

The röntgen (named after the German physicist 1845–1923) is the unit of radiation. It is the amount of X-ray or gamma-ray radiation which produces ions carrying one e.s. unit of either sign in one cubic centimetre of dry air at STP.

The rad (radiation absorption dose) is the unit of absorbed dose and is the energy imparted to matter by an ionizing radiation. It is equivalent to 0.01 joule/kg or 6.242×10^{10} MeV per kilogram of irradiated material. An exposure of 1 röntgen will produce an absorbed dose of 0.869 rad in air.

The rem (röntgen equivalent man) expresses the biological effect of radiation on the human body and is measured in J/kg.

A radiation survey meter with a Geiger–Muller tube detecting the radiation (Fig. 5.5) enables the radioactivity level to be indicated at any point in mR/h. It has a probe connection socket and may be fitted with an audible warning. Other adjustable types give a red-light and audible warning when the radiation exceeds a given value of say 5 mR/h.

Characteristics of gamma-ray sources used in gamma-radiography

Source	Material thickness mm		Half life	Gamma energies MeV	Exposure rate (R/h) (for 1 Ci equivalent activity at 1 m)
	Steel	Light alloys			
cobalt-60	50.0–150.0	150.0–450.0	5.27 years	1.730:1.333	1.3000
iridium-192	12.5–62.5	40.0–190.0	74.00 days	0.206–0.612	0.4800
thulium-170	2.5–12.5	7.5–37.5	128.00 days	0.052:0.084	0.0025
ytterbium-169	2.5–15.0	7.5–45.0	32.00 days	0.008–0.308	0.1250

Only trained and radiation-classified personnel are allowed to operate X-ray and gamma-ray equipment. Generally they wear a small film of the dental type so that when the film is developed each operator knows to how much radiation he or she has been exposed.

The most stringent precautions are taken to protect workers from the harmful effects of radiation and the operation, storage and transport are covered by British Standards, HMSO and Department of the Environment and ISO publications, etc. Students requiring further information should consult the literature on industrial radiography published by the film manufacturers and the suppliers of radioisotopes.

Physical quantity	Unit	SI unit	Conversion
exposure	rontgen (R)	coulomb/kilogram (C/kg)	1 C/kg = 3876 R 1 R = 2.58 × 10⁻⁴ C/kg
activity	becquerel (Bq) 1 Bq = 1/s	curie (Ci)	1 Bq = 2.7 × 10⁻¹¹ Ci 1 Ci = 3.7 × 10¹⁰ Bq
absorbed dose	rad (rad)	gray (Gy) 1 Gy = 1 J/kg	1 Gy = 100 rad 1 rad = 0.01 Gy
equivalent dose	rem (rem)	sievert (Sv) 1 Sv = 1 J/kg	1 Sv = 100 rem 1 rem = 0.01 Sv.

Fig. 5.5. Radiation survey meter.

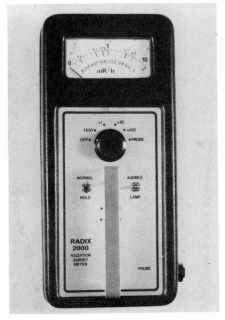

X-ray method

X-rays are produced by an X-ray tube which consists of an evacuated glass bulb with two arms. One arm houses the cathode, a filament which is heated by an electric current as in an electric light bulb, and this heated filament gives off a stream of electrons (negatively charged particles). In the other arm is the anode, which is a metal stem (see Fig. 5.6). By placing a high voltage of the order of 30–500 kV and upwards between anode and cathode the electrons are attracted at high speed to the anode and are focussed into a beam by means of a focussing cup. Fixed in the anode at an angle to the electron beam is the anticathode. This is a dense, high melting point slab of metal such as tungsten, on to which the electron beam impinges and is arrested. The resulting loss of kinetic energy appears as heat and X-rays and the latter emerge from the tube at right angles to its axis (Fig. 5.6). The tube current, which indicates the intensity of flow of the electrons, is in mA and the intensity of the radiation is somewhat proportional to this mA value.

The hardness of X-rays is their penetrating power, which increases with their energy and is inversely proportional to their wavelength. Those with short wavelength are hard rays and those with long wavelength, soft rays. Usually soft radiation has about 20–60 kV, hard rays 150–400 kV, while very hard rays may have over 400 kV on the tube.

Since only part of the kinetic energy (0.1% at 30 kV and 1% at 200 kV and increasing) is converted into radiation, the remaining energy is transformed into heat so that the cathode must be cooled by: (1) radiation, (2) convection or (3) forced circulation by fluid, depending upon the type of

Fig. 5.6. Diagram of X-ray tube.

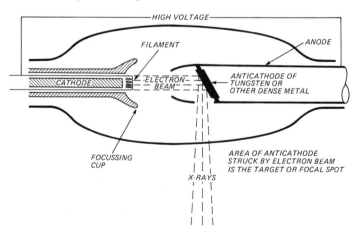

tube. The area of the anticathode must be sufficiently large to avoid overheating or burning.

The rays can penetrate solid substances but, in doing so, a certain proportion of the rays is absorbed and the amount of the absorption depends upon the thickness of the substance and its density. The denser and thicker the substance, the smaller the proportion of X-rays that will get through. X-ray films are made of many layers on a base of cellulose triacetate or polyester, the small silver halide crystals which are sensitive to the rays being suspended in gelatine.

The film is placed in a rigid or flexible cassette with intensifying screens on either side so as to improve the image. The weld or object to be radiographed is placed on the cassette in the path of the rays as shown in fig. 5.7 and after exposure for a short time, depending upon the thickness of the object, the film is developed either manually or automatically. The weld will appear as a light band across the X-ray negative, Fig. 5.8. Any defects in the weld can be seen as dark areas of faults such as blowholes, porosity, etc. Tungsten inclusions as in TIG welding will appear as very light patches, as the tungsten is very dense (Fig. 5.8e). It can be seen that the X-ray film is really a shadowgraph.

Small pipe welds can be X-rayed by directing the rays at an angle to the pipe axis as shown in Fig. 5.9.

To obtain a correctly exposed negative and thus ensure that the smallest defect is visible, image quality indicators (IQI) are used. These may be of the wire type (DIN) with several parallel wires of varying diameter, the sensitivity being the number of the thinnest wire that is just visible on the radiograph. An American method uses small metal plates of aluminium,

Fig. 5.7

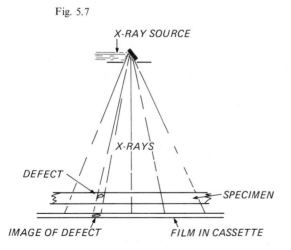

Fig. 5.8 (*a*). Double operator TIG process. Material: Aluminium–magnesium alloy plate ASTM SB20a-5183 (N8) 8 mm thickness. Square butt with 4 mm gap preparation. Current 115 A, TIG process alternating current manual double operator vertical with zirconiated electrodes, welding both sides of the joint simultaneously. Wire diameter 3 mm, type BS 2901 Pt 4 5556A (NG61). Shielding gas argon. The radiograph shows porosity and oxide inclusions probably associated with stop–start region. ASTM = American Society Testing and Materials, AWS = American Welding Society, ASME = American Society of Mechanical Engineers.

Fig. 5.8 (*b*). Material: Stainless steel pipe, 3 mm thickness BS 970 Pt 4 304S12, AISI 304L. Chromium 18%, nickel 10%, carbon 0.03%. Current 120 A, voltage 25 V, manual TIG. Wire diameter 2 mm BS 2901 308S92 AWS 308L. Chromium 19%, nickel 9%, carbon 0.03%. Shielding gas argon. Thoriated electrode. This pipe would normally be purged. In this case the purge is ineffective resulting in uneven penetration profile. The line showing near the D marker is secondary penetration without back purge which has occurred during the filling pass.

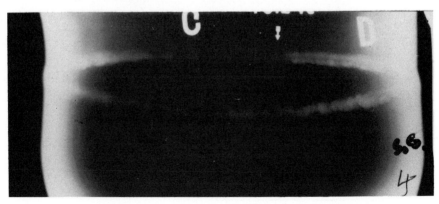

Fig. 5.8 (*c*). Material: Aluminium–magnesium alloy ASTM SB210 5154-0 (N5) 6 mm thickness. Single preparation with permanent backing strip. Current 180 A, alternating current manual TIG. Wire diameter 3 mm type BS 2901 5556A (NG61). Shielding gas argon. Zirconiated electrode. Showing large tungsten inclusion.

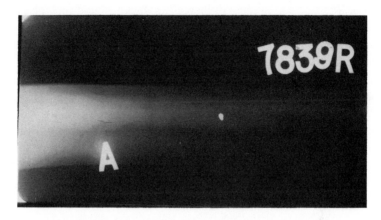

Fig. 5.8 (*d*). Submerged arc weld circumferential seam. Cylinder diameter 2 m. Material: Carbon steel BS 4360 080M50 (43A) thickness 7 mm. Square butt preparation with permanent backing strip. Current 440 A, voltage 32–35 V, electrode +ve; speed of travel 500 mm/min. Wire BS 2901 grade 18, diameter 3.2 mm. Acid fused flux. Weld produced from one side only. The scalloped effect on the radiograph is due to a gap between backing strip and shell resulting in roll under. There is also a large cavity apparent. This radiograph was made with an iridium-192 (gamma-ray) capsule since reduced access prevented use of the X-ray tube on this weld area.

copper, steel, etc., their thickness being usually 2% of the material being radiographed. Small holes, of diameter which are multiples of the plate thickness, are drilled in the plates and the quality of the image is given by the smallest diameter hole visible on the radiograph. British (BWRA) indicators use step wedges of increasing thickness and the holes drilled in each step are the thickness of the step, the last number visible indicating the sensitivity. These can also be used, neglecting the perforations, as ordinary stepped wedges and noting the thinnest visible step (Fig. 5.10).

The flow of electrons within the tube is measured in mA (the cathode current) and the high voltage required between anode and cathode is generally obtained by using the self-rectifying action of the tube with its heated filament. Electron flow is from cathode to anode as long as the anode is kept cool. As its temperature rises during the working of the tube the anode begins to emit electrons, the tube gradually ceases to be self-rectifying and the cathode filament may suffer. For this reason tube heads

Fig. 5.8 (*e*). Submerged arc weld; longitudinal seam in $2\frac{1}{2}$ m diameter pipe. Plate material: Carbon steel BS 4360 080M50 (43A) thickness 7 mm. Square butt preparation. Current 400–430 A, voltage 28–30 V, electrode +ve; speed of travel 500 mm/min. Wire BS 2901 grade 18, diameter 3.2 mm. Acid fused flux. Weld produced from both sides with no back chip.

Fig. 5.9. Radiographing a pipe weld.

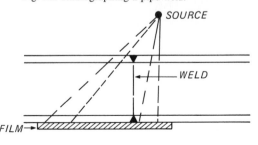

can be fitted with a thermal trip which limits the temperature to about 75 °C and there are connexions for compressed air and water coolants. This type of rectification is half-wave and pulsating, and other means of rectification use semi-conductors with various circuits to avoid the pulsating voltage and to enable the tube to be used for longer periods without overheating.

A typical modern industrial tube unit may be directional or panoramic. The former has a self-rectifying tube, thermal trip, small focal spot, and the tube and transformer within the unit are oil insulated with a cooling system for connexion to compressed air or water. Cathode current is 4.0–8.0 mA and depends upon the tube size. They are available for thicknesses of steel from 25 mm to 75 mm with a kV range of 55 to 300.

The panoramic tube has a similar specification but the tube has a conical target and the 360 ° forward throw unit enables one-shot exposure of pipe welds to be made as the unit will pass into a 230 mm diameter opening. A lead belt with a radiation port is also supplied to screen the radiation during warm up and can be used to convert the tube to directional function by removal of the lead cover. For pipe exposures the film is fastened to the exterior of the pipe with the tube inside the pipe (Fig. 5.11).

The automatic control unit for the tube head works off 110 or 240 V a.c. circuits and includes an automatic build-up of high tension and tube current. This will operate after initial exposure by manual regulation of the mA and kV controls and repeat exposures can be made by pressing the 'in'

Fig. 5.10. Image quality indicator (penetrameter).

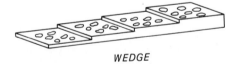

WEDGE

Fig. 5.11. One-shot examination of pipe weld with gamma-ray source or panoramic X-ray tube.

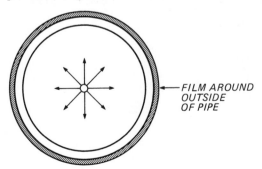

FILM AROUND OUTSIDE OF PIPE

button. The head has also a rotatable D ring which is used for positioning during exposure and for carrying (Fig. 5.12).

Gamma-ray method

Like X-rays, gamma-rays show a shadowgraph on a sensitized film and are interpreted in the same way.

The advantages of radioisotope sources for radiographic purposes are that they need no power supply nor cooling system. Their small focus makes them very suitable for weld inspection in narrow pipes and because some radioisotopes have high powers of penetration, thick specimens can be radiographed at shortened exposure time. They have, however, harder radiation than an X-ray tube so that the image has less contrast and interpretation is more difficult. Also the activity decreases appreciably with those radioisotopes that have a short half life so that their radioactivity depends upon the time, since renewal, and a time–activity curve must be consulted when using them. The radioactivity of the source cannot be varied or adjusted and since it cannot be switched off, it has to be effectively shielded.

The radioactive source is a pellet of a substance encased in a welded stainless steel container about 15 mm long by 5 mm diameter. The pellet is a cylinder of the pure metal cobalt-60 and iridium-192 and a pressed and sintered pellet of thulium dioxide–thulium-170 (Fig. 5.13). These radioactive pellets do not induce radioactivity in the container and the source can be

Fig. 5.12. Positioning of X-ray tubes.

returned, after a certain period depending upon its half life, to the makers to be re-energized in an atomic reactor.

The source must be stored inside a container with a dense radiation shield, usually made of lead, tungsten or even depleted uranium, where it is kept until actually in use. One type has a shutter mechanism for exposure, another type has the source mounted inside the removable portion of the shield, which can be detached and used like a torch so that the radiation appears forwards, away from the operator's body and shielded in the backwards direction. This type is useful for most work, including pipe welds.

The third type shown in Fig. 5.14*a* and *b* has the radioisotope mounted on a flexible cable and contained within a shielded container. It can be pushed along the guide tube by remote control and can be positioned in otherwise awkward places. With this type, positioning and source changing is easily performed. Pipeline crawlers for various diameter pipes are used, carrying the radioisotope and enabling it to be positioned in the pipe centre to give a radial beam of radiation when exposed, Fig. 5.11. The film is placed around the outside of the pipe enabling the radio-inspection at that point to be performed with one exposure. The crawler can be battery operated and travels on wheels with forward, reverse, expose and stop controls, the positioning within the pipes being controlled to a few millimetres accuracy.

Examples

Source	Pellet size (mm) diamond length	Maximum equivalent activity (Ci)	Approximate absorber dose in air at 1 m (mGy/h)
cobalt-60	2 × 2	10–15	110–170
iridium-192	2 × 2	50	210

Fig. 5.13. Typical capsule for cobalt-60 and iridium-192. Material: stainless steel.

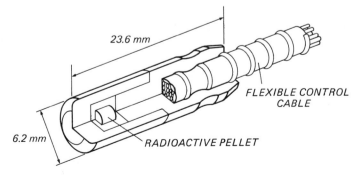

23.6 mm

6.2 mm

FLEXIBLE CONTROL CABLE

RADIOACTIVE PELLET

Fig. 5.14 (*a*). Remote control exposure container (shutterless).

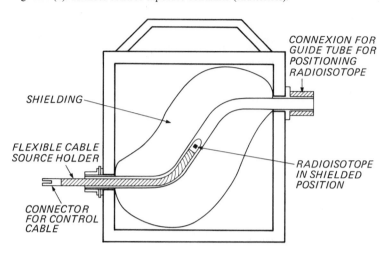

CONNEXION FOR
GUIDE TUBE FOR
POSITIONING
RADIOISOTOPE

SHIELDING

FLEXIBLE CABLE
SOURCE HOLDER

RADIOISOTOPE
IN SHIELDED
POSITION

CONNECTOR
FOR CONTROL
CABLE

Fig. 5.14 (*b*)

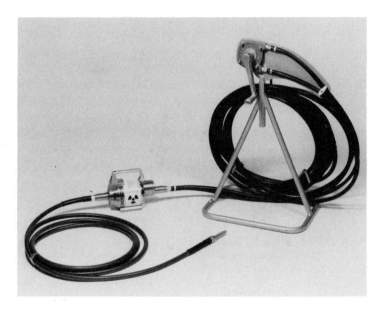

Ultrasonic testing

Ultrasonic testing employs waves above the frequency limit of human audibility and usually in the frequency range 0.6 to 5 MHz. A pulse consisting of a number of these waves is projected into the specimen under test. If a flaw exists in the specimen an echo is reflected from it and from the type of echo the kind of flaw that exists can be deduced.

The equipment comprises an electrical unit which generates the electrical oscillations, a cathode ray tube on which pulse and echo can be seen, and probes which introduce the waves into the specimen and receive the echo. The electrical oscillations are converted into ultrasonic waves in a transducer which consists of a piezo-electric element mounted in a perspex block to form the probe, which, in use, has its one face pressed against the surface of the material under test. When a pulse is injected into the specimen a signal is made on the cathode-ray tube, Fig. 5.15. The echo from a flaw is received by another probe, converted to an electrical e.m.f. (which may vary from microvolts to several volts) by the transducer and is applied to the cathode ray tube on which it can be seen as a signal displaced along the time axis of the tube from the original pulse (Fig. 5.16a).

The first applications of ultrasonics to flaw detection employed longitudinal waves projected into the specimen at right angles to the surface (Fig. 5.16b). This presented problems because it meant that the weld surface had to be dressed smooth before examination, and more often than not the way in which the flaw oriented, as for example lack of penetration, made detection difficult with this type of flaw. The type of wave used to overcome these disadvantages is one which is introduced into the specimen at some distance from the welded joint and at an angle to the surface (e.g. 20°) and

Fig. 5.15

is known as a shear wave. The frequency of the waves (usually 2.5 and 1.5 MHz for butt welds), the angle of incidence of the beam, the type of surface and the grain size, all affect the intensity of the echo which is adjustable by means of a sensitivity control. The reference standard on which the sensitivity of the instrument can be checked consists of a steel block 300 × 150 × 12.7 mm thickness with a 1.6 mm hole drilled centrally and perpendicularly to the largest face, 50.8 mm from one end. Echoes are obtained from the hole after 1, 2 or 3 traverses of the plate (Fig. 5.16c) and from the amplitude of the echo the intensity from a hole of known size can be checked.

Three types of probe are available:

(1) A single probe which acts as both transmitter and receiver, the same piezo-electric elements transmitting the pulse and receiving the echo. The design of the probe is complicated in order to prevent reflections within the perspex block confusing the echo.

(2) The twin transmitter–receiver probe in which transmitter and receiver are mounted together either side by side or one in front of the other but are quite separate electrically and ultrasonically so that there is no trouble with interference with the echo. This type is the most popular (Fig. 5.17a).

(3) The separate transmitter and receiver each used independently (two-handed operation) (Fig. 5.17b).

To make a 'length scan' of the weld the transmitter–receiver unit is moved continuously along a line parallel to the welded seam so that all points of the whole area of the welded joint are covered by the scanning beam, and care must be exercised that by the use of too high a spread of the beam, double echoes are not obtained from a single flaw. It is evident that varying the distance from the weld to the probe varies the depth at which the main axis of the beam crosses the welded joint and moving the probe at

Fig. 5.16

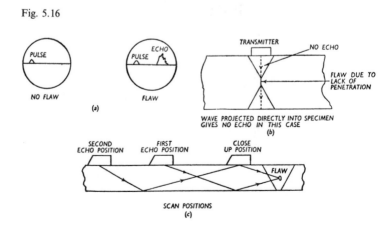

right angles to the line is thus known as depth scan. A spherical flaw will have no directional characteristics and a wave falling upon its centre will be reflected along the incident path, the amplitude of the echo depending upon the size of the flaw. Cylindrical flaws behave in the same way but in the case of a narrow planar flaw it is evident that optimum echo will be received when the crack is at right angles to the wave and there will be no echo when the crack lies along the wave, but if the probe is moved to the first echo position the crack is no longer lying along the beam.

The probes must make good contact with the specimen and on slightly curved surfaces a thin film of oil is used to improve the contact. On surfaces with greater curvature, as for example when investigating circumferential welds on drums, curved probes are used.

We have only considered the essential points of ultrasonic testing and it must be emphasized that there is a considerable amount of theory involved in the relationships between the distance of the transmitter from the weld and the beam angle, etc., and that a great amount of practice is required to interpret correctly the echoes received and from them decide the nature and position of the flaw. See also BS 3923, *Methods for ultrasonic examination of welds*, Parts 1 and 2.

Application of load

An illustration of this method is furnished by the hydraulic test on boilers. Water is pumped into the welded boiler under test (the safety valve if fitted having been clamped shut) to a pressure usually $1\frac{1}{2}$ to 2 times the working pressure. Should a fault develop in a joint, the hydraulic pressure rapidly falls without danger to persons near, such as there would have been if compressed air or steam had been used.

In the same way, partial compressive or tensile loads may be applied to any welded structure to observe its behaviour. The method adopted will, of course, depend on the nature of the work under test.

Fig. 5.17

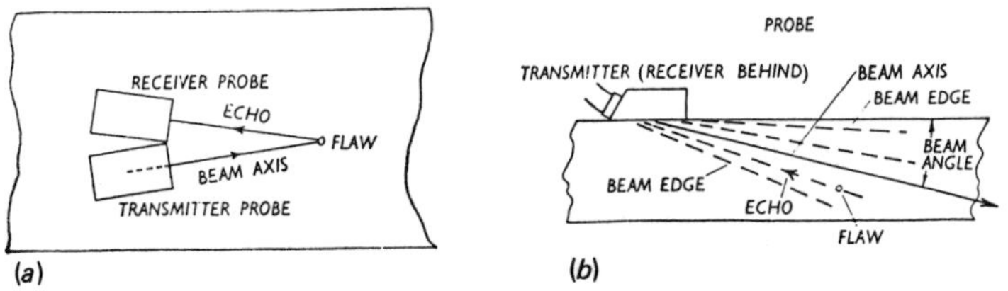

Destructive tests

These may be divided as follows:
(1) Tests capable of being performed in the workshop.
(2) Laboratory tests, which may be divided as follows: (1) microscopic and macroscopic, (2) chemical, analytical and corrosive, (3) mechanical.

Workshop tests

(The student is advised to study BS 1295, *Tests for use in the training of welders*, which gives standard workshop tests for butt and fillet welds in plates, bars and pipes.)

These are usually used to break open the weld in the vice for visual inspection. When operators are first learning to weld, this method is very useful, because as a rule the weld contains many defects and, when broken open, these can quickly be pointed out. Little time is thus lost in finding out the faults and rectifying them. As the welding technique of the beginner improves, however, this test becomes of much less value. Obviously much will depend on the actual position of the specimen in the vice, whether held on the joint or just below it. Also on the hammering, whether heavy erratic blows are used or a medium-weight, even hammering is given. In addition, if the weld metal is stronger than the parent metal, fracture may occur in the parent metal and thus the weld itself has hardly been tested. We can make sure that the specimen will break in the weld and afford us opportunity for examination by making a nick with a hacksaw as shown on each end of the weld, having previously filed or ground the ends square (see Fig. 5.18).

Another useful method for determining the ductility of the weld is to bend the welded specimen in the vice through 180° with an even bending force. Any cracks appearing on the weld face will indicate lack of ductility. A better method of conducting this test will be described later (see Fig. 5.19).

A useful workshop test, for use in the case in which the welded parts have

Fig. 5.18. Specimen for 'nick bend' or 'nick break' test.

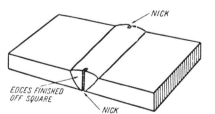

to be heated up or even forged after welding, consists of actually forging a test specimen after welding. It is always advisable to apply the tests given later also, such as tensile, in order to obtain the ultimate strength of the weld.

Workshop tests are very limited, and their chief advantage is the little time taken to perform them. They are useful during training of welders, but little knowledge of the weld can be gained from them. The visual method, as previously explained, is a valuable addition to the workshop methods given above.

Microscopic and macroscopic tests

Microscopic tests. The use of the microscope is very important in determining the actual structure of the weld and parent metal. When a polished section of the weld is observed with the eye, it will look completely homogeneous if no blowholes or entrapped slag are present. On the other hand, if a section is broken open, as in the nick bend test, it may be found that there is a definite crystal-like structure. Since, however, this type of section may have broken at the weakest line, we must take a section across any desired part of the weld in order to have a typical example to examine. Specimens to be microscopically examined are best cut by means of a hacksaw. Any application of heat, as for example with gas cutting, may destroy part of the structure which it is desired to examine. If this specimen was polished by means of abrasives in the usual commercial way, when observed under the microscope it would be found to be covered with a multitude of scratches.

The best method of preparation is to grind carefully the face of the specimen after cutting on a water-cooled slow-running fine grinding wheel of large diameter, care being taken to obtain a flat face. Polishing can then be continued by hand, using finer abrasives, finally polishing by the polishing wheel, using rouge or aluminium oxide as the abrasive.

Fig. 5.19

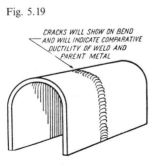

In order to bring out the structure of the section of metal clearly, the surface must now be 'etched'. This consists of coating it with a chemical which will eat away and dissolve the metal. Since the section is a definite structure consisting of composite parts, some are more easily dissolved than others, and thus the etching liquid will bring up the pattern of the structure very clearly when observed under the microscope.

The etching liquids employed depend on the metal of the specimen. For iron and steel, a 1 or 2% solution of nitric acid in alcohol, or picric acid in alcohol, is used. For copper, either ammonium persulphate or ferric chloride acidified with hydrochloric acid. For aluminium and aluminium alloys, either caustic soda or dilute hydrofluoric acid and nitric acid.

Most of the microphotographs in this book were prepared by etching with the 2% nitric acid solution in alcohol.

The length of time for which the etching liquid remains on the metal depends on the detail and the magnification required. After etching is complete, the liquid is washed off the surface of the specimen to prevent further action. For example, if steel etched with picric acid is to be examined at 100 diameters, etching could be carried out from 25 to 35 seconds, giving a clear, well-cut image. If this, however, was observed under the high-power glass of 1000 diameters, it would be found that picric acid had eaten deeply into the surface, and the definition and result would be extremely poor. Thus, for high magnification, the etching would only need carrying out for 5–10 seconds. Naturally, however, the time will vary entirely with the etching liquid used, the power of magnification and the detail required.

When the section is prepared in this way and the whole crystal structure is visible, the exact metallic condition of the weld can be examined, together with that of the surrounding parent metal. For example, examination of microphotographs of steel at 150 to 200 diameters will indicate the size of the grain, the arrangement of pearlite and ferrite. Increasing magnification to 1000 diameters will indicate the presence of oxides or nitrides, oxides being shown up as fine cracks between the crystals (producing weakness) (Fig. 2.19), and iron nitrides as needle-like crystals (producing brittleness) (Fig. 2.21). From this, the metallurgist can tell the suitability of the weld metal, how well the structure compares with that of the parent metal, and its probable strength. This study or test plays an important part in the manufacture of new types of welding rods. Microphotographs of varying magnification are used in various parts of this book to illustrate the structures referred to.

Macroscopic tests. This method consists, as before, of preparing a cross-section of the weld by polishing and etching. It is then examined either with a low-power microscope magnifying 3 to 20 diameters or even with a magnifying glass. This will show up any cracks, entrapped slag, pin-size blow or gas holes, and will also indicate any coarse structure present (Fig. 5.20).

Fig. 5.20
(*a*) Fillet weld, single-run each side, with good penetration and no undercut.
(*b*) Fillet weld, two runs on one side (× 3.) Note good fusion.

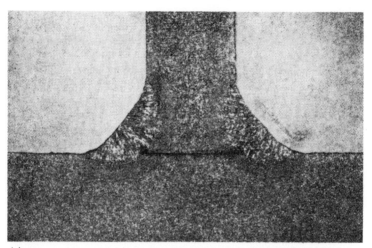

(a)

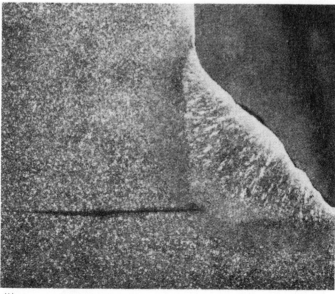

(b)

The etching fluids most suitable for macroscopic examination are:

Steel and iron: 10% iodine, 20% potassium iodine and 70% distilled water; 10 to 20% nitric acid in water; 8% cuprous ammonium chloride in water.

Copper: 25% solution of nitric acid in water; ammonium hydrate; nitric acid in alcohol.

Brass and bronze: 25% solution of nitric acid.

Aluminium and aluminium alloys: 10% solution of hydrofluoric acid in water.

The macrographic examination of welds can easily be undertaken in the workshop, using a hand magnifying glass, and the degree of polish required is not so high as for microscopic examination. The microscope, however, will obviously bring out defects and crystal structures which will not be apparent in the macrograph.

Sulphur prints. This is an easy method by which the presence of sulphur, sulphides and other impurities can be detected in steel. It is not suitable for non-ferrous metals or high-alloy steels.

The principle of sulphur printing is that a dilute acid such as sulphuric will attack sulphur and sulphides, liberating a gas, hydrogen sulphide (H_2S), which will stain or darken bromide or gaslight photographic paper.

To make a sulphur print, the specimen is first prepared by filing or machining and then by rubbing by hand or machine to obtain a scratch-free surface (O grade emery). A piece of photographic paper is soaked in dilute sulphuric acid for about 3 to 4 minutes and then after excess acid has been sponged off, the paper is laid carefully on the prepared surface of the steel specimen and pressed down perfectly flat on its surface. It is left on the specimen for about 4 to 5 minutes, the edge of the paper being lifted at intervals to ascertain how it is staining. After removing it, the paper is treated as in photographic printing, namely rinsed, then immersed in a 20% hypo solution for few minutes and then again thoroughly washed. The darker the stains on the paper the higher the sulphur content.

Chemical tests

Analytical tests are used to determine the chemical composition of the weld metal. From its composition, the physical properties of the metal can be foretold. The addition of manganese increases the toughness of steel, uranium increases its tensile strength, and these are indicated fully in the chapter on metallurgy.

Corrosive tests are used to foretell the behaviour of the weld metal under conditions that would be met with in years of service.

The action of acids and alkalis, present in the atmosphere of large industrial areas and which may have a marked effect on the life of the welded joints, can be observed, the effect in the laboratory being concentrated so as to be equal in a few days to years of normal exposure. From these tests, the most suitable type of weld metal is indicated. The following examples will serve as illustrations.

Along the sea coast, greatest corrosion takes place to those metal parts which are subject to the action both of the salt water and the atmosphere, that is, the areas between high and low tide; for example, oil-rig and landing-stage supports and caissons, and railings and structures exposed to the sea spray. By dipping welded specimens alternately in and out of a concentrated brine solution corrosion effects equal to years of exposure are produced.

Suppose it is required to compare the resistance to acid or alkaline corrosion of plates welded together with different types of electrodes. The specimens are polished and marked and then photographed. They are then rotated in a weak acid or alkaline solution. The specimens are photographed at given intervals and the degree of corrosion measured in each case. From the results it is evident which electrode will give the best resistance to this particular type of corrosion.

In the chemical industry, tanks are required for the storing of corrosive chemicals. It is essential that the welded joints should be just as proof against corrosion as the metal of the tank itself. Corrosive tests undertaken as above in the laboratory will indicate this, and will enable a correct weld metal to be produced, giving proof against the corrosion.

Evidently, then, these tests are specialized, in that they reproduce as nearly as possible, in the laboratory, conditions to which the weld is subjected.

Mechanical tests
These may be classified as follows:
(1) Tensile.
(2) Bending.
(3) Impact: Charpy and Izod.
(4) Hardness: Brinell, Rockwell, Vickers Diamond Pyramid (Hardness Vickers HV) and Scleroscope.
(5) Fatigue: Haigh and Wöhler.
(6) Cracking: Reeve.

Tensile test
As stated previously, a given specimen will resist being pulled out in the direction of its length and up to a point (the yield point) will remain

elastic, that is, if the load is removed it will recover its original dimensions. If loaded beyond the yield point or elastic limit the deformation becomes permanent.

Preparations of specimens. In order to tensile test a welded joint, specimens are cut from a welded seam and one specimen from the plate itself. This latter will give the strength of the parent metal plate. The specimens are machined or filed so as to have all the edges square, and the face can be left with the weld built up or machined flat, depending on the test required. It is usual, in addition, to cut specimens for bend testing from the same plate, and these are usually cut alternately with the tensile specimens, as shown in Figs. 5.21 and 5.24.

If the elongation is required, it is usual to machine the specimen flat on all faces and to make two punch marks 50 mm apart on each side of the weld, as shown in Fig. 5.22. Fig. 5.23*a* and *b* shows two specimens prepared for tensile test.

Preparation of all-weld metal specimens

Two steel plates approximately $200 \times 100 \times 20$ mm thick are prepared with one face at an angle of $80°$ as in Fig. 5.25. The plates are set about 16 mm apart on a steel backing strip about 10 mm thick and are welded in position. The groove is built up with the weld metal under test and with a top reinforcement of about 3 mm. The welded portion is then cut out (thermally) along a line about 20 mm each side of the weld line. A

Fig. 5.21

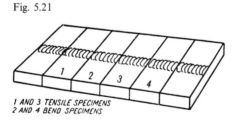

I AND 3 TENSILE SPECIMENS
2 AND 4 BEND SPECIMENS

Fig. 5.22

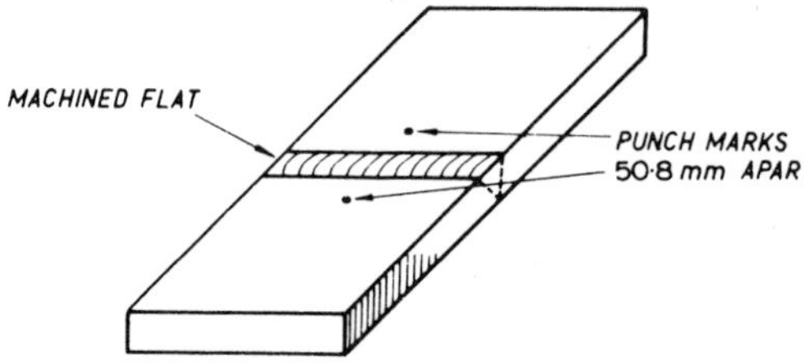

MACHINED FLAT

PUNCH MARKS
50·8 mm APART

Fig. 5.23. The preparation of two specimens for tensile test with cross-sectional areas of 75 mm² and 150 mm².

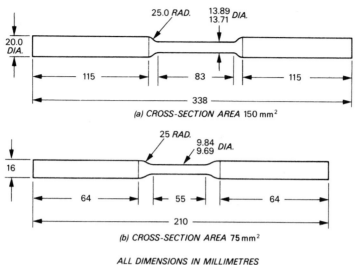

(a) CROSS-SECTION AREA 150 mm²

(b) CROSS-SECTION AREA 75 mm²

ALL DIMENSIONS IN MILLIMETRES

Fig. 5.24. Preparations of specimens for test.

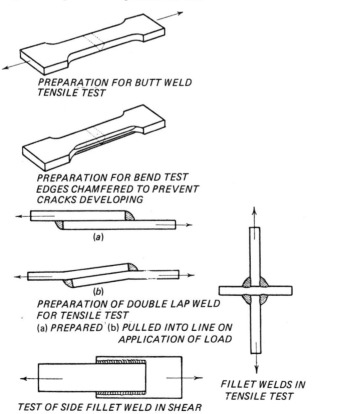

PREPARATION FOR BUTT WELD TENSILE TEST

PREPARATION FOR BEND TEST EDGES CHAMFERED TO PREVENT CRACKS DEVELOPING

(a)

(b)

PREPARATION OF DOUBLE LAP WELD FOR TENSILE TEST
(a) PREPARED (b) PULLED INTO LINE ON APPLICATION OF LOAD

FILLET WELDS IN TENSILE TEST

TEST OF SIDE FILLET WELD IN SHEAR

tensile specimen is prepared from the all-weld metal as in Fig. 5.26 and a Charpy specimen as in Fig. 5.34.

The specimen is prepared from deposit well away from the parent plate as there will be effects of dilution on the two or three initial layers.

Testing machines

Present-day testing machines are available in a variety of designs suitable for specialized testing. A typical universal machine for tests in tension, cold bend, compression, double shear, transverse, cupping and punching has four ranges, 0–50, 100, 250 and 500 kN (50 tonf).

It comprises an hydraulic pumping unit (Fig. 5.27) with a multi-piston pump of variable displacement enabling the movement of the straining rams to be closely controlled. Hydraulic pressure is applied to the pistons of the rams *R* which move the cross beam *H* to which the straining wedge box is attached. The specimen under test is gripped between this and an upper wedge box connected to a series of lever balances *A*, *B* and *C*, the movement of which when load is applied is indicated on the figure. These balances or beams are mounted on hardened steel knife edges and are of deep section to

Fig. 5.25. Preparation of an all-weld metal test piece.

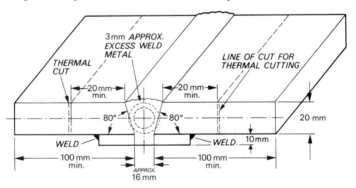

VIEW IN DIRECTION OF ARROW X

Fig. 5.26. An all-weld tensile test specimen.

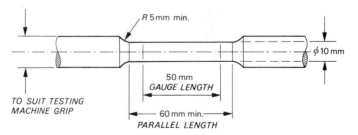

prevent deflexion. The balance arm *D* is attached at one end to the weighing spring *W* and the other end actuates the dial pointer, which moves over a scale gradually in kN. An additional spring *T* helps to keep the knife edges in contact and an oil dash pot *O* acts as a damper. The arm *C* has four fulcra, any of which can be selected by movement of a hand-operated cam depending upon the range of the test required, and at the same time the range selected is indicated on the dial.

Tensile test of a welded joint. It will be evident that a tensile test on a welded joint is not quite similar to a test on a homogeneous bar, and the following considerations will make this clear. The steel weld metal may be strong, yet brittle and hard. When tested in the machine, the specimen would most probably break outside the weld, in the parent metal, whereas in service, due to its brittleness, failure might easily occur in the weld itself. The result of this test gives the tensile strength of the bar itself and indicates that the weld is sound. It does not indicate any other condition.

If the weld metal is softer than the parent metal, when tested the weld metal itself will yield, and fracture will probably occur in the weld. Because of this, the elongation of the specimen will be small, since the parent bar will have only stretched a small amount, and this would lead to the belief that the metal had little elasticity. Quite on the contrary, however, the weld metal may have elongated by a considerable amount, yet because of its small size in comparison to the length of the specimen the actual elongation observed is small. Great care must therefore be taken to study carefully the results and to interpret them correctly, bearing in mind the properties which it is required to test.

Fig. 5.27. Testing machine layout.

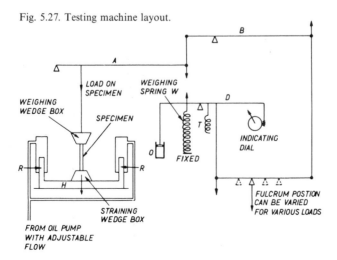

A tensile test on an all-weld metal specimen prepared as previously explained indicates the strength and ductility of the metal in its deposited condition and is a valuable test.

A very useful form of test is that known as the longitudinal test. In this test the weld runs along the length of the test piece. As the load is applied, if the weld metal is ductile, it will elongate with the parent metal and is placed in the machine so that the load is applied longitudinally to help to share the load. If, on the other hand, it is brittle, it will not elongate with the parent metal but will crack. Should the parent metal be of good quality and structure, the cracks will be confined to the weld metal mostly and will merely increase in width. If the parent metal is not of such good quality, the cracks will extend into the parent metal and breakage will occur with little elongation of the specimen. This test therefore indicates the quality of the parent metal as well as that of the weld metal.

Torsion test. This test is useful to test the uniformity of work turned out by welders. A weld is made between steel plates V'd in the usual manner, and a cylindrical bar is turned out of the deposited metal. This specimen is then gripped firmly at one end, while the other end is rotated in a chuck or other similar device, until breakage occurs. The degree of twist which occurs before breakage will depend upon the type of metal under test.

Bend test (for ductility of a specimen)

In this test the bar is prepared by chamfering the edges to prevent cracking (if it is of rectangular section), and is then supported on two edges and loaded at the centre (Fig. 5.28a).

Fig. 5.28. Uniform bar supported at S_1 and S_2 and loaded at the centre. (*a*) Beam deflected elastically. (*b*) Bottom layer stress diagram for elastic deflexion, i.e. extension of bottom surface layers. (*c*) Beam deformed or yielded plastically between X_1 and X_2 but with elastic deflexion between X_2S_2 and X_1S_1. (*d*) Bottom layer stress diagram when yield commences.

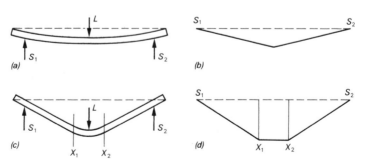

As the load is applied the bar first bends elastically, and in this state if the load is removed it would regain its original shape. On increasing the load a point will be reached when the fibres of the beam at the centre are no longer elastically deformed, i.e., they have reached their yield point, and the bar deforms plastically at the centre (Fig. 5.28*c*).

Further increase of load causes yielding to occur farther and farther from the centre, while at the same time the stress at the centre increases. Ultimately, when maximum stress is reached, fracture of the bar will occur. If this maximum stress is not reached, fracture of the bar will not occur for any angle of bend. The method determining the ductility of the bar from the above test is as follows. Lines are scribed on the machined or polished face of the specimen parallel and equidistant to each other across a width of about 150–250 mm. As the load is applied, the increase in distance between the scribed lines is measured, and this increase is plotted vertically against the actual position of the lines horizontally. When the bar deforms elastically the result is a triangle (Fig. 5.28*b*), termed the stress diagram, since it represents graphically the stress at these points. The stress diagram is shown for plastic deformation of the bar in Fig. 5.28*d*.

Now consider the test applied to a welded joint and let the weld be placed in position under the applied load. There are now two different metals to be considered, since the weld metal might have quite different properties from those of the parent metal (Fig. 5.29).

If the load is applied and the yield point of the weld metal is greater than that of the parent bar, plastic yield or bend will occur in the bar, and as the load is increased the bar bends plastically, as in Fig. 5.29*b*. During this bending, if the yield point of the weld metal is reached, the weld metal will flow or yield somewhat, but in any case most of the bend is taken by the bar. If the yield point of the weld metal is not reached, then all the bend will be taken by the bar.

If, however, the yield point of the weld metal is lower than that of the bar, the weld metal will first bend plastically and will continue to do so, plastic deformation occurring long before the yield point of the bar is reached (Fig. 5.29*d*). On such a small area as the weld metal has, therefore, fracture will occur in the weld metal at a small angle of bend. The stress diagrams given in Fig. 5.29*c* and *e* indicate where the greatest elongations of the fibres occur in each case.

Since the weld metal is almost always harder or softer than the parent metal the bending will not occur, therefore, equally in the weld and in the parent metal, and as a result the chief value of this test is to determine whether any flaws exist in the weld. Otherwise its value as a test of ductility of a welded specimen is very limited.

If the weld is placed so that its face is under the central applied load, fracture will occur at the root of the weld if the penetration is imperfect (Fig. 5.30).

Impact tests

We have seen when discussing notch brittleness in steel that localized plastic flow of a notch may cause cracking and that the transition from ductile to brittle state is affected by temperature, strain rate and the occurrence of notches. It should be noted that as there is no ductile–brittle transition with aluminium, impact tests are performed to a lesser degree with aluminium than with steel. Serrations, tool marks, changes of section and other discontinuities on the surface of metals that are met with in service reduce their endurance so that the term 'notch sensitivity' is applied to the degree to which these discontinuities reduce the mechanical properties. This is an important consideration in welding because, for example, any reduction in section due to undercut along the toes of butt welds and in the vertical plate in HV fillets reduces the mechanical properties of the structure.

To determine the notch brittleness (or notch toughness), impact tests are

Fig. 5.29. (*a*) Bar with inserted wedge of weld metal having mechanical properties differing from that of the bar itself, supported at S_1 and S_2 and loaded at the centre. (*b*) Load applied. Yield point of weld not reached. Bend taking place in the bar. (*c*) Bottom layer stress diagram. Length of XY represents extension at point X. (*d*) Load applied. Yield point of *bar* not reached. Bend taking place in the weld. (*e*) Bottom layer stress.

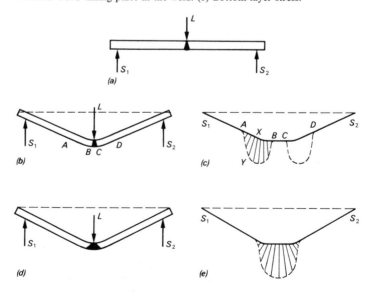

performed on specimens prepared with a notch of precise width, depth and shape, and the resistance which the specimen offers to breaking at the notch when hit by a striker moving at a given velocity and having a given energy is a measure of the notch brittleness.

The two main tests, Charpy and Izod, employ a swinging pendulum to which a slave pointer is attached. This moves over a scale calibrated in joules as the pendulum swings and stays at the impact value of the test, being afterwards reset by hand. The pendulum tup (or bob) incorporating the striker hits the notched specimen at a given velocity and with a given energy (measured in joules). If no specimen were present the pendulum would swing unhindered to the zero position on the scale, but since energy is lost in breaking the specimen the pointer will take up a position at say x joules on the scale. This is the impact value for the specimen at the particular temperature and represents the energy lost by the pendulum in breaking the specimen.

Impact tests are being increasingly used at sub-zero temperatures in order to give indications and possibilities of brittle fracture. Diethyl ether and liquid nitrogen are used to obtain temperatures down to $-196\,°C$ using a copper–constantan thermo-couple for temperature measurement.

Charpy and Izod machines

Machines can be Charpy and Izod combined or Charpy only or Izod only and can be manually or pneumatically operated. In manual machines the pendulum is lifted physically to the start position where it is

Fig. 5.30. A typical transverse bend test. Upper and lower surfaces are ground or machined flat. The specimen is aboud 30 mm wide. The bending should be through an angle of 180° over a former with a diameter three times that of the plate thickness. Test should be made with: (1) the weld face in tension, and (2) the root of the weld in tension.

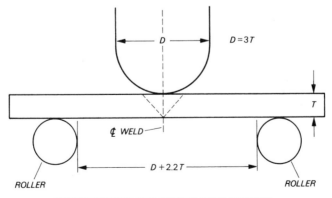

TRANSVERSE BEND TEST ON BUTT WELD SPECIMEN

held in position by a release box which has a self-setting catch. There is a pendulum release lever and a safety lock lever which prevents accidental release of the pendulum. Pneumatically operated machines operate in a similar manner to the manual machines except that the pendulum is lifted under power to the Charpy or Izod start position and can be set for automatic release (Fig. 5.31a and b).

Machine capacities are 0–150 J (striker velocity 3–4 m/s) for the Izod machine and 0–300 J (striker velocity 5–5.5 m/s) with an optional 0–150 J for the Charpy machine. On the combined machines, tups and strikers are changed for the different tests and gauges are provided for the Charpy machine to check that the striker hits the specimen centrally between the anvils.

Charpy test

This test may be either with a V or a U section notch, the specimen and notch sizes being shown in Fig. 5.32. the V notch test is becoming increasingly used in Britain and is the test required for impact values in BS 639 – *Covered electrodes for the MMA welding of carbon and carbon–manganese steels.*

Fig. 5.31. (*a*) The Charpy machine; 150 and 300 joules.

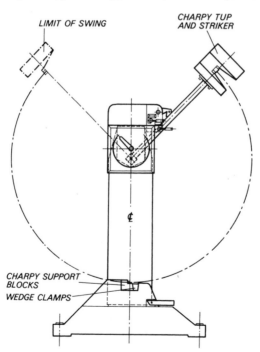

Fig. 5.31a and b indicates the method of operation of the machine. The specimen is supported squarely at its two ends by machine supports, the notch being centrally placed by means of small tongs (Fig. 5.33). The pendulum is raised to the test height and the pointer indicates 300 J on the scale. A hand lever is operated, the pendulum swings and the striker hits the specimen exactly on the side behind the notch. Energy is absorbed in fracturing the specimen and the pointer swings to say x joules on the scale, this being the Charpy value at the particular temperature on the 300 J scale for either V or U notch, whichever was chosen. Fig. 5.34a shows the

Fig. 5.31 (b)

Fig. 5.32. Preparation of V and U notches.

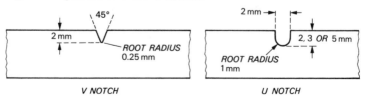

V NOTCH   U NOTCH

SPECIMENS 55 mm LONG AND 10 mm SQUARE SECTION

NOTCHES CENTRALLY PLACED AND AT RIGHT ANGLES TO THE LONGITUDINAL AXIS

preparation of an all-weld metal test piece and Fig. 5.35*b* shows the preparation for impact testing a weld. (See also BS 131 Pt 2, *Charpy* V *test* and Pt 3, *Charpy* U *notch test*.)

Izod test

This test is performed on a specimen with a V notch and of dimensions as in Fig, 5.35*a*. The specimen is mounted vertically in a groove in the vice wedge block assembly, which is tightened by handwheel.

Fig. 5.33. Striker and specimen for the Charpy test.

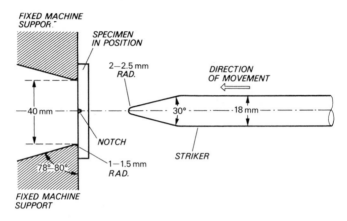

Fig. 5.34. (*a*) An all-weld metal test piece for Charpy test. (*b*) The preparation for impact testing.

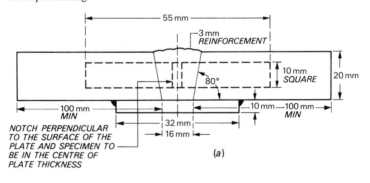

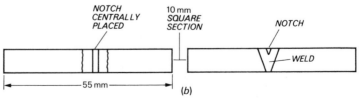

The V notch is located facing the striker and with the base of the V exactly in line with the top edge of the vice, and is lined up with a small hand jig. The striker hits the specimen at the striking height 22 mm above the V notch (Fig. 5.35*b*).

To operate the machine the pendulum is raised to the 150 J position and upon release swings as in the Charpy test and the pointer indicates the impact value in joules. Fig. 5.36 shows a combination machine with Izod release box in position and set for the Izod test and also with the Charpy release box in position. (See also BS 131 Part 1.)

Hardness tests

These are useful to indicate the resistance of the metal to wear and abrasion, and give a rough indication of the weldability of alloy steels. If parts, such as tramway crossings, dredger bucket lips, plough shares or steel gear wheels, have been reinforced or built up, it is essential to know the degree of hardness obtained in the deposit. This can be determined by portable hardness testers of the following types.

The chief methods of testing are: (*a*) Brinell, (*b*) Rockwell, (*c*) Vickers Diamond Pyramid, (*d*) Scleroscope.

Fig. 5.35. The Izod test (single notch).

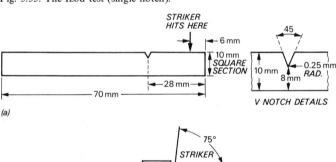

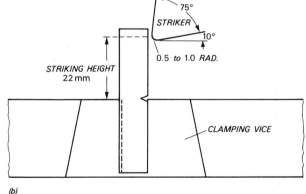

The ***Brinell test*** consists in forcing a hardened steel ball, 10 mm diameter, hydraulically into the surface under test. The area in square millimetres of the indentation (calculated from the diameter measured by a microscope) made by the ball, is divided into the pressure in kilograms, and the result is the Brinell hardness number or figure. Figure 5.37 shows an indentation of 4.2 mm diameter when measured on the microscope scale.

For example: if the load was 3000 kg, and the area of indentation 10 mm², Brinell number equals 3000 divided by 10, which is 300.

The Brinell figure can be calculated from the following:

$$\text{Brinell figure} = \frac{-P}{\dfrac{\pi D}{2}\,(D - \sqrt{(D^2 - d^2)})}$$

where P = load in kg, D = diameter of ball in mm, d = diameter of indentation in mm.

The tensile strength of mild steel in N/mm² is approximately 3.4 times the Brinell hardness value.

Fig. 5.36. Machine for the Izod test; 0–150 joules.

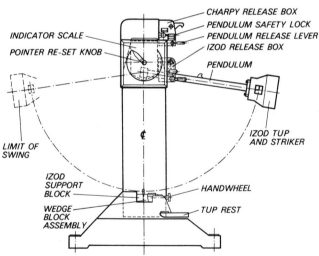

Fig. 5.37. Indention made by Brinell ball in hard surface and measured by microscope scale.

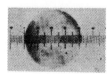

Evidently the ball must be harder than the metal under test or the ball itself will deform and as a result it is used only up to a figure of about 500. See also BS 240 Pt 1 and 2.

Rockwell hardness test

There are three standard indenters: a diamond cone with an included angle of 120° and with the tip rounded to a radius of 0.2 mm; a 1.6 mm diameter hardened steel ball; and a 3.2 mm diameter hardened steel ball. The diamond cone or steel ball is first pressed into a clean surface under test with a load of F_1 kgf to a point B, distant AB above a reference line at A (Fig. 5.38a). A further load F_2 is now applied making a total load of $F_1 + F_2$ kgf and the indenter is pushed further into the surface to a point C, distant AC above A. The major load is then released leaving only the initial load F_1, and there is some recovery to the point D, distant AD above A. Then BD represents the permanent depth of indentation due to the additional load. This indentation is automatically measured on the dial of the machine and indicates the Rockwell hardness HR, the number having an added letter indicating the scale of hardness used.

In each case in the following typical scales the initial load is 10 kgf (F_1).

Scale A, additional load 50 kgf, total load 60 kgf, diamond indenter.

Scale B, additional load 90 kgf, total load 100 kgf, 1.6 mm diameter steel ball indenter.

Scale C, additional load 140 kgf, total load 150 kgf, diamond indenter.

BD represents the permanent indentation due to the additional load. The reference plane at A represents the zero of the particular hardness scale, AB having a constant value of 100 units for the diamond indenter and 130 for the steel ball indenter. To use a testing machine the particular indenter in

Fig. 5.38. (*a*) *BD* represents the permanent indentation due to the additional load. The reference plane at *A* represents the zero of the particular hardness scale, *AB* having a constant value of 100 units for the diamond indenter and 130 for the steel ball indenter.

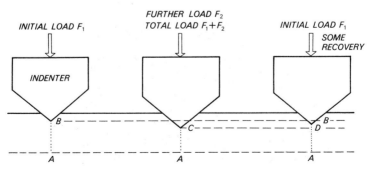

use is fitted to the machine and the scale selected. Loads are applied automatically, the hardness number appearing on the dial (Fig. 5.38*b*) so that routine hardness tests are easily and quickly performed. See also BS 891 Pt 1 and 2.

Vickers hardness test (BS 427)

The Vickers hardness test is similar to the Brinell test but uses a square base right pyramidal diamond with an angle of 136° between opposite faces as the penetrator. The two diagonals d_1 and d_2 mm of the identation are measured and the average calculated. The Hardness Vickers is obtained thus:

$$HV = 1.854 \frac{F}{d^2},$$

where F = load in kgf and d is the diameter of the diagonal (or the average of the diameters) (Fig. 5.39). To avoid having to perform this calculation for each reading, the diameter is obtained from an ocular reading fitted to

Fig. 5.38. (*b*) Direct reading hardness testing machine.

the measuring microscope and the HV is obtained from the ocular reading with the use of a table.

The pressure, which is applied for a short time, can be varied from 1 to 120 kgf according to the hardness of the specimen under test. The Brinell number and the HV number are practically the same up to 500, the Brinell number being slightly lower (see table).

The ***Scleroscope test*** consists of allowing a hard steel cylinder, called the hammer, having a pointed end, to fall from a certain height onto the surface under test. The height to which it will rebound will depend upon the hardness of the surface, and the rebound figure is taken as the hardness figure. The fall is about 250 mm, giving with a hard steel surface a rebound height of about 150 mm, this being 90 to 100 on the Scleroscope scale.

Fatigue test

If a specimen is subjected to a continuously alternating set of push and pull forces operating for long periods, the specimen may fail due to

Brinell number	Hardness Vickers	Rockwell	
		Scale A	Scale C
59	100	43	
235	240		20
380	400		40
430	460		45
460	500		48
535	600		54
595	700		58
	800		62
	1000		68

Fig. 5.39

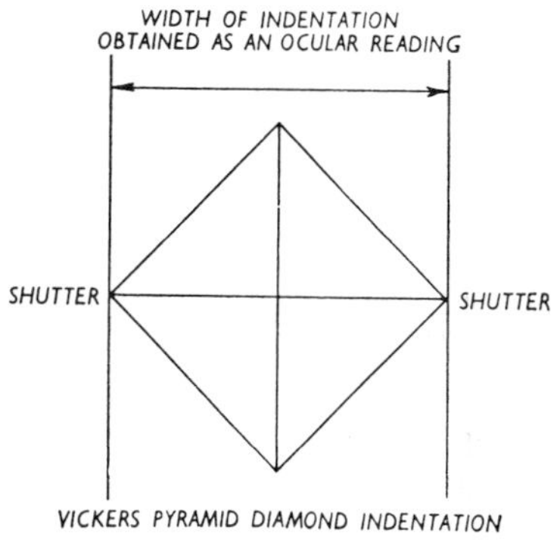

WIDTH OF INDENTATION
OBTAINED AS AN OCULAR READING

SHUTTER

SHUTTER

VICKERS PYRAMID DIAMOND INDENTATION

fatigue of the molecules, and the magnitude of the force under which it may fall will be much less than its maximum tensile or compressive strength. The forces applied rise to a maximum tension, decrease to zero, rise to a maximum compression and decrease again to zero. This is termed a cycle of operations and may be written 0–maximum tension–0–maximum compression–0 and so on. Fatigue tests are based on this phenomenon exhibited by metals.

In the *Haigh tests*, a soft-iron core or armature vibrates between the poles of an electro-magnet carrying alternating current, and is connected to the specimen under test. As alternating current is passed through the coil, the armature vibrates at the frequency of the supply (usually 50 Hz) between the poles, and the welded specimen is thus subjected to alternating push–pull forces at this frequency. The alternating current, and therefore the force on the armature, rises as above: 0–maximum in one direction–0 minimum–0–maximum in opposite direction–0; this being, as before, one cycle of operations.

The drawback to this test is that at 3000 reversals per minute an endurance test of 10 000 000 reversals would take about 56 hours and a complete endurance test will take many days.

The latest type of electromagnetic fatigue tester gives approximately 17 000 reversals per second, and thus the required tests can be performed in a fraction of the time and the machine automatically shuts off when failure occurs.

The pick-up which causes the vibration is controlled from an oscillator and non-magnetic metals can also be tested.

In the *Wöhler test*, the specimen is gripped at one end in a device like a chuck and the load is applied at the other end by fixing it to a bearing, as shown in Fig. 5.40. When the chuck rotates at speed, the specimen is continuously under an alternating tension and compression, tension when the face of the weld is uppermost, as shown, and compression when it is below. If the load applied is large, the specimen may be pulled out of

Fig. 5.40. Illustrating principle of Wöhler test.

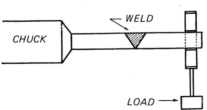

balance. These out-of-balance forces then increase the forces on the specimen, and we are unable to tell the load under which the weld failed. This can be overcome, however, by a slight modification of the machine having a bearing at each side of the joint under test and the load applied between the bearings, but the test remains the same. In conducting a fatigue test, a certain load is placed on the specimen, and this produces a certain stress. Suppose the stress produced is 140 N/mm²; this stress varies from zero to 140 N/mm² tensile stress, then back to zero and to 140 N/mm² compressive stress and back to zero. This is a complete cycle. Low speed reversals of an applied force may be applied hydraulically over long periods to find the resistance to fatigue of a welded structure.

Fatigue tests are extremely useful for observing the resistance to fatigue of welded shafts, cranks and other rotating parts, which are subjected to varying alternating loads. They also provide a method of comparing the resistance to fatigue of solid drop forged and welded fabricated components.

Cracking (Reeve) test

This is used in the study of the hardening and cracking of welds and is of especial value in ascertaining the weldability of low-alloy structural steels and high tensile steels, which as before mentioned are prone to harden and develop cracks on cooling. A 150 mm square plate of the metal to be welded is placed on another larger plate of the same metal and the two are firmly secured to a heavy bed plate, 50 mm or more in thickness, by means of bolts as shown in Fig. 5.41.

Edges *a*, *b* and *c* are then welded with any selected electrode, thus firmly welding the two plates together, and they are then allowed to cool off. Edge *d* is the one on which the test run is to be deposited using the electrode under test, and evidently, since the two plates are completely restrained in movement, any tendency to crack on cooling will show in the weld on the edge *d*.

Fig. 5.41. Reeve test.

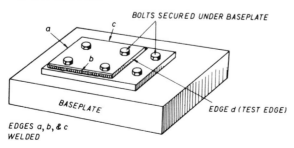

After cooling, the bolts are removed and the weld examined by previously described methods for cracks. Sections can then be sawn off from the plate, the hardness of the weld tested at various points, and sections etched and examined microscopically.

It can be seen from this outline of available tests for welds that the particular test chosen will depend entirely upon the type of welded joint and the conditions under which it is to operate. These conditions will govern the tests which must be applied to indicate the way in which the weld will behave under actual service conditions.

Erichsen test (cupping)

This is used for determining the suitability of a metal for deep drawing and pressing. A punch with a rounded head is pushed into the surface of the metal, the depth and appearance of the indentation before cracking occurs being an indication of the suitability for drawing and pressing.

6

Engineering drawing and welding symbols

Engineering drawing

The principal method usually adopted in the making of machine drawings is known as orthographic projection.

Suppose the part under construction is shown in Fig 6.1*a*. This 'picture' is known as an isometric view. It is of small use to the engineer, since it is difficult to include on it all the details and dimensions required, especially those on the back of the picture, which is hidden.

Imagine that around the object a box is constructed (O being the corner farthest from the observer) having the sides Ox, Oy, Oz all at right angles to each other. The plane or surface of the box bounded by Ox and Oy is the

Fig. 6.1. (*a*)

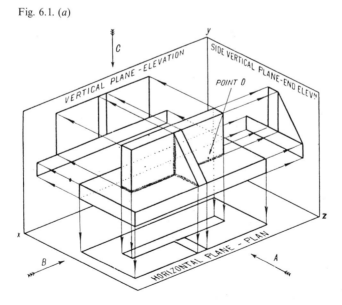

vertical plane, indicated by VP; that bounded by Oy, Oz is the side vertical plane, SVP, and that bounded by Ox and Oz, the horizontal plane, HP, these three planes being the three sides of the box farthest from the observer. Lines are projected, as shown, onto these planes from the object under consideration, and the view obtained when looking at the object in the direction of the arrow A. The end elevation is the view obtained by projection on to the side vertical plane, while the plan is the view obtained by projection on to the horizontal plane. The arrow B shows the direction in which the object is viewed for the side elevation and C the direction for the plan.

Now imagine the sides VP, SVP and HP opened out on their axes Ox, Oy and Oz. The three projections will then be disposed in position, as shown in Fig. 6.1b, i.e. the end elevation is to the *right* of the elevation and the plan is *below* the elevation.

On these three projections, which are those used by the engineer, almost all the details required during manufacture can be included, and hence they are of the greatest importance.

This method is known as First Angle Projection, the projection lines being shown in Fig. 6.1a. It is now rarely used.

A second method, called Third Angle Projection, is extensively used nowadays, and can be understood by reference to Fig. 6.2a.

From this it will be seen that the corner of the box O is chosen to be that nearest the observer, and the three planes are those sides of the box also nearest to the observer, the part under consideration being seen through

Fig. 6.1. (b)

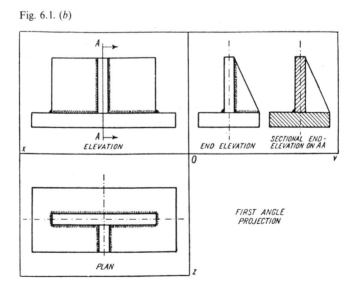

these planes of projection. The elevation is again that view formed by projection on to the vertical plane Ox, Oy, the end elevation that formed by projection on the side vertical plane Oy, Oz and the plan that formed by projection on the horizontal plane Ox, Oz. Owing, however, to the change in the axes when they are unfolded, the projections are disposed differently, the plan now being *above* the elevation and the side elevation being to the *left* of the elevation (Fig. 6.2b). (N.B. The object is viewed in the same direction as previously, as indicated by the arrows.)

Fig. 6.2. (*a*)

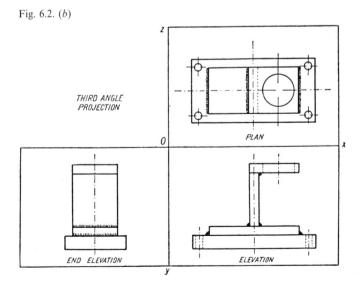

Fig. 6.2. (*b*)

By noting the above difference between the two methods the welder can immediately tell which method has been used. Sometimes a combination of these two methods may be encountered but need not be considered here.

Welding symbols*

A weld is indicated on a drawing by (1) a symbol (Fig. 6.3*a*) and (2) an arrow connected at an angle to a reference line usually drawn parallel to the bottom of the drawing (Fig. 6.3*b*). The side of the joint on which the arrow is placed is known as the 'arrow side' to differentiate it from the 'other side' (Fig. 6.3*c*). If the weld symbol is placed *below* the reference line, the weld face is on the *arrow-side* of the joint, while if the symbol is above the reference line the weld face is on the other side of the joint (Fig. 6.3*c*). Symbols on both sides of the reference line indicate welds to be made on both sides of the joint, while if the symbol is across the reference line the weld is within the plane of the joint. A circle where arrow line meets reference line indicates that it should be a peripheral (all round) weld, while a blacked in flag at this point denotes an 'on site' weld (Fig. 6.3*d*). Intermittent runs of welding are indicated by figures denoting the welded portions, and figures in brackets the non-welded portions, after the symbol (Fig. 6.3*e*). A figure before the symbol for a fillet weld indicates the leg length. If the design throat thickness is to be included, the leg length is prefixed with the letter '*b*' and the throat thickness with the letter '*a*'. Unequal leg lengths have a × sign separating the dimensions (Fig. 6.3*f*). A fork at the end of the reference line with a number within it indicates the welding process to be employed (e.g. 131 is MIG, see table on p. 286), while a circle at this point containing the letters NTD indicates that non-destructive testing is required (Fig. 6.3*g*). Weld profiles, flat (or flush), convex and concave profiles are shown as supplementary symbols in Fig. 6.3*h*.

Figure 6.4*a* gives the elementary symbols and 6.4*b* typical uses of them. The student should study BS 499, Part 2, 1980, *Symbols for welding*, for a complete account of this subject.

A table showing the numerical indication of processes complying with International Standard ISO 4063 appears on p. 286.

---

* BS 499, Part 2, 1980, brings the standard in line with ISO recommendations.

Fig. 6.3

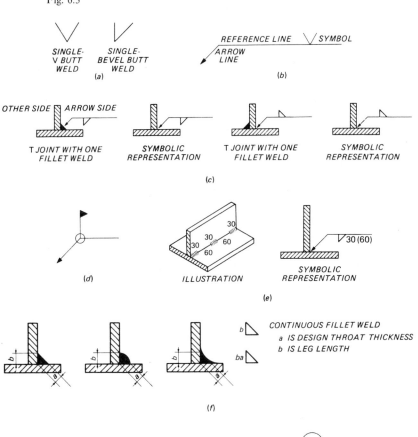

SINGLE-
V BUTT
WELD

SINGLE-
BEVEL BUTT
WELD

(a)

REFERENCE LINE   SYMBOL

ARROW
LINE

(b)

OTHER SIDE   ARROW SIDE

T JOINT WITH ONE
FILLET WELD

SYMBOLIC
REPRESENTATION

T JOINT WITH ONE
FILLET WELD

SYMBOLIC
REPRESENTATION

(c)

(d)

ILLUSTRATION

SYMBOLIC
REPRESENTATION

(e)

CONTINUOUS FILLET WELD
a IS DESIGN THROAT THICKNESS
b IS LEG LENGTH

(f)

PROCESS   NON-DESTRUCTIVE TESTING

(g)

Supplementary symbols

SHAPE	SYMBOL
FLAT (USUALLY FINISHED FLUSH)	⎯
CONVEX	⌒
CONCAVE	⌣

Examples of supplementary symbols

DESCRIPTION	ILLUSTRATION	SYMBOL
FLAT (FLUSH) SINGLE-V BUTT WELD		
CONVEX DOUBLE-V BUTT WELD		
CONCAVE FILLET WELD		
FLAT (FLUSH) SINGLE-V BUTT WELD WITH FLAT (FLUSH) BACKING RUN		

(h)

Numerical indication of process (*BS 499, Part 2, 1980*)*

No.	Process	No.	Process
1	**Arc welding**	43	Forge welding
11	Metal-arc welding without gas protection	44	Welding by high mechanical energy
		441	Explosive welding
111	Metal-arc welding with covered electrode	45	Diffusion welding
		47	Gas pressure welding
112	Gravity arc welding with covered electrode	48	Cold welding
113	Bare wire metal-arc welding	**7**	**Other welding processes**
114	Flux cored metal-arc welding	71	Thermit welding
115	Coated wire metal-arc welding	72	Electroslag welding
118	Firecracker welding	73	Electrogas welding
12	Submerged arc welding	74	Induction welding
121	Submerged arc welding with wire electrode	75	Light radiation welding
		751	Laser welding
122	Submerged arc welding with strip electrode	752	Arc image welding
		753	Infrared welding
13	Gas shielded metal-arc welding	76	Electron beam welding
131	MIG welding	77	Percussion welding
135	MAG welding: metal-arc welding with non-inert gas shield	78	Stud welding
		781	Arc stud welding
136	Flux cored metal-arc welding with non-inert gas shield	782	Resistance stud welding
14	Gas-shielded welding with non-consumable electrode	**9**	**Brazing, soldering and braze welding**
		91	Brazing
141	TIG welding	911	Infrared brazing
149	Atomic-hydrogen welding	912	Flame brazing
15	Plasma arc welding	913	Furnace brazing
18	Other arc welding processes	914	Dip brazing
181	Carbon arc welding	915	Salt bath brazing
185	Rotating arc welding	916	Induction brazing
		917	Ultrasonic brazing
2	**Resistance welding**	918	Resistance brazing
21	Spot welding	919	Diffusion brazing
22	Seam welding	923	Friction brazing
221	Lap seam welding	924	Vacuum brazing
225	Seam welding with strip	93	Other brazing processes
23	Projection welding	94	Soldering
24	Flash welding	941	Infrared soldering
25	Resistance butt welding	942	Flame soldering
29	Other resistance welding processes	943	Furnace soldering
291	HF resistance welding	944	Dip soldering
		945	Salt bath soldering
3	**Gas welding**	946	Induction soldering
31	Oxy-fuel gas welding	947	Ultrasonic soldering
311	Oxy-acetylene welding	948	Resistance soldering
312	Oxy-propane welding	949	Diffusion soldering
313	Oxy-hydrogen welding	951	Flow soldering
32	Air fuel gas welding	952	Soldering with soldering iron
321	Air-acetylene welding	953	Friction soldering
322	Air-propane welding	954	Vacuum soldering
		96	Other soldering processes
4	**Solid phase welding; Pressure welding**	97	Braze welding
41	Ultrasonic welding	971	Gas braze welding
42	Friction welding	972	Arc braze welding

* This table complies with International Standard ISO 4063.

Fig. 6.4. (*a*) Elementary welding symbols (BS 499, Part 2, 1980).

DESCRIPTION	SECTIONAL REPRESENTATION	SYMBOL
1. BUTT WELD BETWEEN FLANGED PLATES (FLANGES MELTED DOWN COMPLETELY)		 ⊐⊏
2. SQUARE BUTT WELD	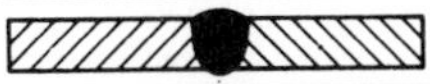	‖
3. SINGLE-V BUTT WELD		∨
4. SINGLE-BEVEL BUTT WELD		�
5. SINGLE-V BUTT WELD WITH BROAD ROOT FACE		Y
6. SINGLE-BEVEL BUTT WELD WITH BROAD ROOT FACE		⋎
7. SINGLE-U BUTT WELD		Y
8. SINGLE-J BUTT WELD		⋃
9. BACKING OR SEALING RUN		⌒
10. FILLET WELD	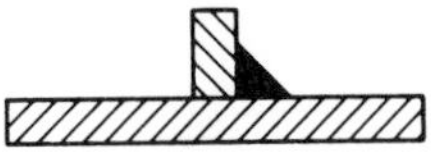	△
11. PLUG WELD (CIRCULAR OR ELONGATED HOLE, COMPLETELY FILLED)	ILLUSTRATION 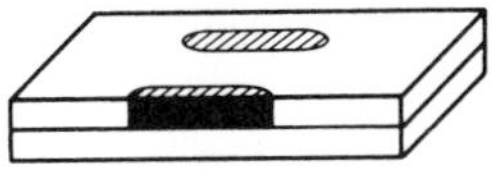	⊓
12. SPOT WELD (RESISTANCE OR ARC WELDING) OR PROJECTION WELD	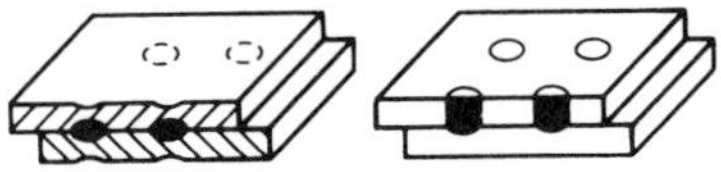 (a) RESISTANCE (b) ARC	○
13. SEAM WELD	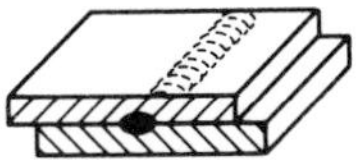	 ⊖

288

Fig. 6.4. (*b*) Examples of uses of symbols (BS 499, Part 2, 1980).

DESCRIPTION SYMBOL	GRAPHIC REPRESENTATION	SYMBOLIC REPRESENTATION
SINGLE-V BUTT WELD $\vee$	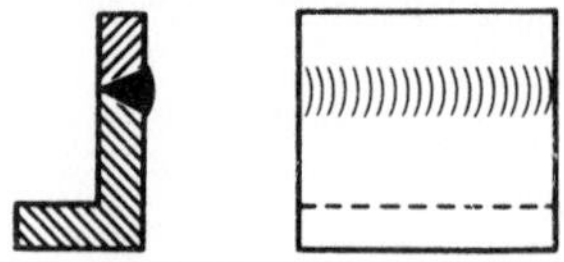	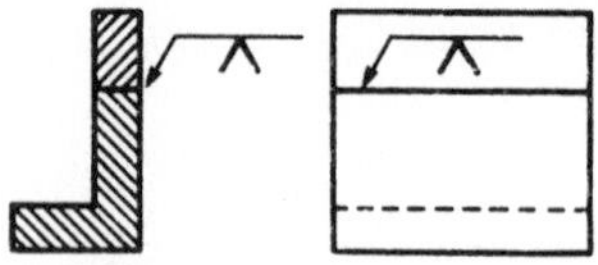
SINGLE-V BUTT WELD $\vee$ AND BACKING RUN	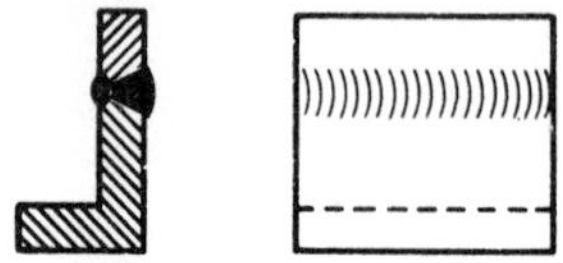	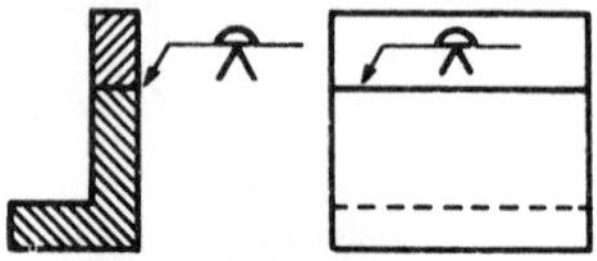
FILLET WELD	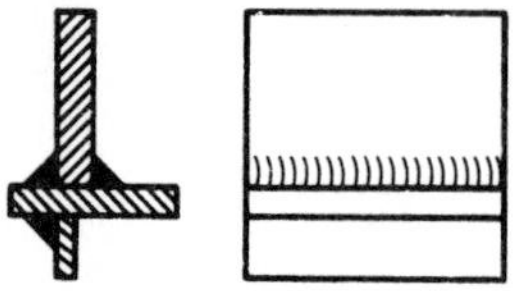	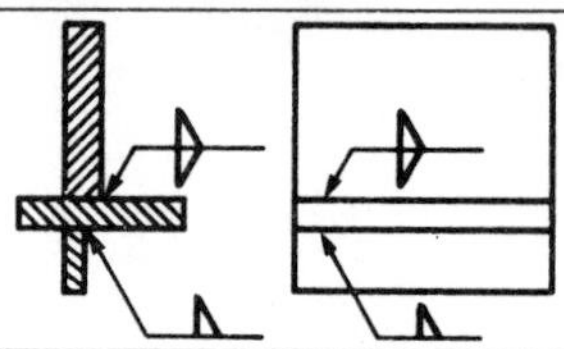
SINGLE-BEVEL BUTT WELD WITH FILLET WELD SUPERIMPOSED	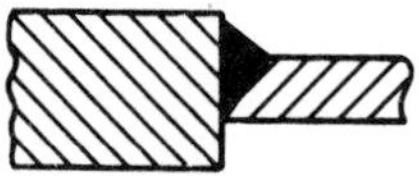	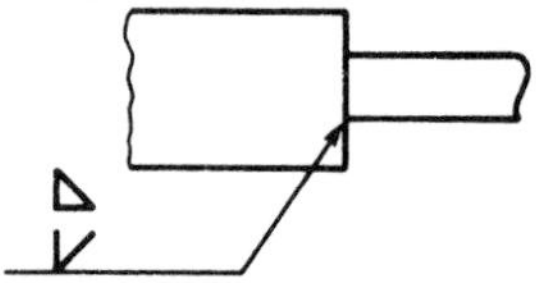
SQUARE BUTT WELD 	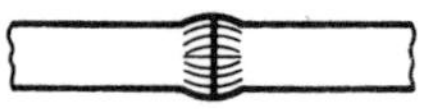	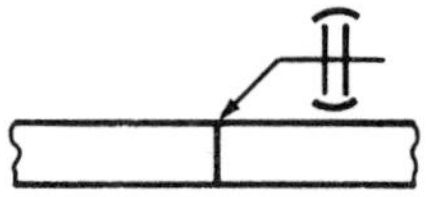
STAGGERED INTERMITTENT FILLET WELD	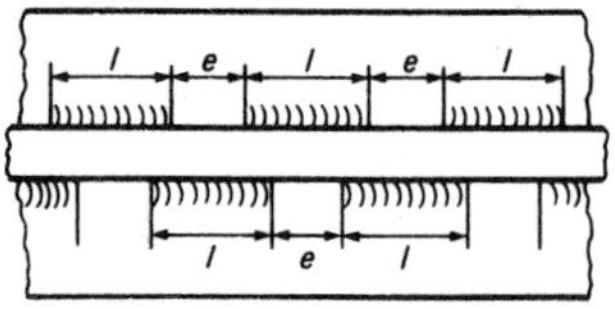	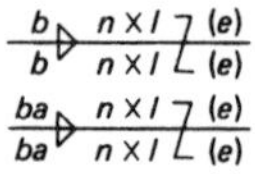 $\dfrac{b}{b} \searrow n \times l \urcorner (e)$ $\dfrac{b}{b} \nearrow n \times l \llcorner (e)$ $\dfrac{ba}{ba} \searrow n \times l \urcorner (e)$ $\dfrac{ba}{ba} \nearrow n \times l \llcorner (e)$ *a* IS THE DESIGN THROAT THICKNESS *b* IS THE LEG LENGTH *e* IS THE DISTANCE BETWEEN ADJACENT WELD ELEMENTS *l* IS THE LENGTH OF THE WELD (WITHOUT END CRATERS) *n* IS THE NUMBER OF WELD ELEMENTS

See Vol. 2, appendix 1, for American symbols.

Appendices

1 Tables of elements and conversions

Elements: their symbols, atomic weights and melting points.

Element	Symbol	Atomic weight	Melting point (°C)
Actinium	Ac	227	—
Aluminium	Al	26.97	658.7
Americium	Am	241	—
Antimony	Sb	121.77	630
Argon	Ar	39.94	− 188
Arsenic	As	74.96	850
Astatine	At	211	—
Barium	Ba	137.36	850
Berkelium	Bk	245	—
Beryllium	Be	9.02	1280
Bismuth	Bi	209.00	271
Boron	B	10.82	2200–2500
Bromine	Br	79.91	− 7.3
Cadmium	Cd	112.41	320.9
Caesium	Cs	132.81	26
Calcium	Ca	40.07	810.0
Californium	Cf	246	—
Carbon	C	12.00	3600
Cerium	Ce	140.13	635
Chlorine	Cl	35.45	− 101.5
Chromium	Cr	52.01	1615
Cobalt	Co	58.94	1480
Copper	Cu	63.57	1083
Curium	Cm	242	—
Dysprosium	Dy	162.5	—
Erbium	Er	167.64	—
Europium	Eu	152	—
Fluorine	F	19.0	− 223
Francium	Fa	223	—
Gadolinium	Gd	157.26	—

289

Elements: their symbols, atomic weights and melting points (contd.)

Element	Symbol	Atomic weight	Melting point (°C)
Gallium	Ga	69.72	30.1
Germanium	Ge	72.60	958
Gold	Au	197.2	1063
Hafnium	Hf	179	2200
Helium	He	4.00	−272
Holmium	Ho	165	—
Hydrogen	H	1.0078	−259
Indium	In	114.8	155
Iodine	I	126.932	113.5
Iridium	Ir	193.1	2350
Iron	Fe	55.84	1530
Krypton	Kr	83.7	−169
Lanthanum	La	138.90	810
Lead	Pb	207.22	327.4
Lithium	Li	6.94	186
Lutecium	Lu	175	—
Magnesium	Mg	24.32	651
Manganese	Mn	54.93	1230
Mercury	Hg	200.61	−38.87
Molybdenum	Mo	96	2620
Neodymium	Nd	144.27	840
Neon	Ne	20.18	−253
Neptunium	Np	237	—
Nickel	Ni	58.69	1452
Niobium (Columbium)	Nb(Cb)	92.9	1950
Nitrogen	N	14.008	−210
Osmium	Os	190.8	2700
Oxygen	O	16.000	−218
Palladium	Pd	106.7	1549
Phosphorus	P	30.98	44
Platinum	Pt	195.23	1755
Plutonium	Pn	239	—
Polonium	Po	210	—
Potassium	K	39.1	62.3
Praseodymium	Pr	140.92	940
Promethium	Pm	147	—
Protactinium	Pa	231	—
Radon	Rn	222	−71
Radium	Ra	226.1	700
Rhenium	Re	186	3167
Rhodium	Rh	102.91	1950
Rubidium	Rb	85.44	38
Ruthenium	Ru	101.7	2450
Samarium	Sm	150.43	1300–1400
Scandium	Sc	45.10	1200
Selenium	Se	78.96	217–220

Elements: their symbols, atomic weights and melting points (contd.)

Element	Symbol	Atomic weight	Melting point (°C)
Silicon	Si	28.06	1420
Silver	Ag	107.88	960.5
Sodium	Na	22.997	97.5
Strontium	Sr	87.63	800
Sulphur	S	32.06	112.8
Tantalum	Ta	181.5	2900
Technetium	Tc	99	—
Tellurium	Te	127.5	452
Terbium	Tb	159.2	—
Thallium	Tl	204.39	302
Thorium	Th	232.12	1700
Tin	Sn	118.70	231.9
Thulium	Tm	169.4	—
Titanium	Ti	47.9	1800
Tungsten	W	184.0	3400
Uranium	U	238.14	1850
Vanadium	V	50.96	1720
Xenon	Xe	131.3	− 140
Ytterbium	Yb	173.6	1800
Yttrium	Y	88.92	1490
Zinc	Zn	65.38	419.4
Zirconium	Zr	91.22	1700

Gauge table. Imperial standard

No.	Size (in)	Size (mm)
0	0.324	8.229
1	0.300	7.620
2	0.276	7.010
3	0.252	6.401
4	0.232	5.893
5	0.212	5.385
6	0.192	4.877
7	0.176	4.470
8	0.160	4.064
9	0.144	3.658
10	0.128 approx. $\frac{1}{8}$ in	3.251
11	0.116	2.946
12	0.104	2.642
13	0.092	2.337
14	0.080	2.032
15	0.072	1.829
16	0.064 approx. $\frac{1}{16}$ in	1.626
17	0.056	1.422

Gauge table. Imperial standard (contd.)

No.	Size (in)	Size (mm)
18	0.048	1.219
19	0.040	1.016
20	0.036	0.914
21	0.032 approx. $\frac{1}{32}$ in	0.813
22	0.028	0.711
23	0.024	0.610
24	0.022	0.559
25	0.020	0.508
26	0.018	0.457
27	0.0164	0.4166
28	0.0148	0.3759
29	0.0136	0.3454
30	0.0124	0.315

Millimetres to inches

mm	0	1	2	3	in 4	5	6	7	8	9
0	——	0.03937	0.07874	0.11811	0.15748	0.19685	0.23622	0.27559	0.31496	0.35433
10	0.39370	0.43307	0.47244	0.51181	0.55118	0.59055	0.62992	0.66929	0.70866	0.74803
20	0.78740	0.82677	0.86614	0.90551	0.94488	0.98425	1.02362	1.06299	1.10236	1.14173
30	1.18110	1.22047	1.25984	1.29921	1.33858	1.37795	1.41732	1.45669	1.49606	1.53543
40	1.57480	1.61417	1.65354	1.69291	1.73228	1.77165	1.81102	1.85039	1.88976	1.92913
50	1.96850	2.00787	2.04724	2.08661	2.12598	2.16535	2.20472	2.24409	2.28346	2.32283
60	2.36220	2.40157	2.44094	2.48031	2.51969	2.55906	2.59843	2.63780	2.67717	2.71654
70	2.75591	2.79528	2.83465	2.87402	2.91339	2.95276	2.99213	3.03150	3.07087	3.11024
80	3.14961	3.18898	3.22835	3.26772	3.30709	3.34646	3.38573	3.42520	3.46457	3.50394
90	3.54331	3.58268	3.62205	3.66142	3.70079	3.74016	3.77953	3.81890	3.85827	3.89764
100	3.93701	3.97638	4.01575	4.05512	4.09449	4.13386	4.17323	4.21260	4.25197	4.29134
110	4.33071	4.37008	4.40945	4.44882	4.48819	4.52756	4.56693	4.60630	4.64567	4.68504
120	4.72441	4.76378	4.80315	4.84252	4.88189	4.92146	4.96063	5.00000	5.03937	5.07874
130	5.11811	5.15748	5.19685	5.23622	5.27559	5.31496	5.35433	5.39370	5.43307	5.47244
140	5.51181	5.55118	5.59055	5.62992	5.66929	5.70866	5.74803	5.78740	5.82677	5.86614
150	5.90551	5.94488	5.98425	6.02362	6.06229	6.10236	6.14173	6.18110	6.22047	6.25984
160	6.29921	6.33858	6.37795	6.41732	6.45669	6.49606	6.53543	6.57480	6.61417	6.65354
170	6.69291	6.73228	6.77165	6.81102	6.85039	6.88976	6.92913	6.96850	7.00787	7.04724
180	7.08661	7.12598	7.16535	7.20472	7.24409	7.28346	7.32283	7.36220	7.40157	7.44094
190	7.48031	7.51969	7.55906	7.59843	7.63780	7.67717	7.71654	7.75591	7.79528	7.83465
200	7.87402	7.91339	7.95276	7.99213	8.03150	8.07087	8.11024	8.14961	8.18898	8.22835
210	8.26772	8.30709	8.34646	8.38583	8.42520	8.46457	8.50394	8.54331	8.58268	8.62205
220	8.66142	8.70079	8.74016	8.77953	8.81890	8.85827	8.89764	8.93701	8.97638	9.01575
230	9.05512	9.09449	9.13386	9.17323	9.21260	9.25197	9.29134	9.33071	9.37008	9.40945
240	9.44882	9.48819	9.52756	9.56693	9.60630	9.64567	9.68504	9.72441	9.76378	9.80315

Thousandths of an inch to millimetres

Mils*	0	1	2	3	mm 4	5	6	7	8	9
0	——	0.0254	0.0508	0.0762	0.1016	0.1270	0.1524	0.1778	0.2032	0.2286
10	0.2540	0.2794	0.3048	0.3302	0.3556	0.3810	0.4064	0.4318	0.4572	0.4826
20	0.5080	0.5334	0.5588	0.5842	0.6096	0.6350	0.6604	0.6858	0.7112	0.7366
30	0.7620	0.7874	0.8128	0.8382	0.8636	0.8890	0.9144	0.9398	0.9652	0.9906
40	1.0160	1.0414	1.0668	1.0922	1.1176	1.1430	1.1684	1.1938	1.2192	1.2446
50	1.2700	1.2954	1.3208	1.3462	1.3716	1.3970	1.4224	1.4478	1.4732	1.4986
60	1.5240	1.5494	1.5748	1.6002	1.6256	1.6510	1.6764	1.7018	1.7272	1.7526
70	1.7780	1.8034	1.8288	1.8542	1.8796	1.9050	1.9304	1.9558	1.9812	2.0066
80	2.0320	2.0574	2.0828	2.1082	2.1336	2.1590	2.1844	2.2098	2.2352	2.2606
90	2.2860	2.3114	2.3368	2.3622	2.3876	2.4130	2.4384	2.4638	2.4892	2.5146
100	2.5400	2.5654	2.5908	2.6162	2.6416	2.6670	2.6924	2.7178	2.7432	2.7686
110	2.7940	2.8194	2.8448	2.8702	2.8956	2.9210	2.9464	2.9718	2.9972	3.0226
120	3.0480	3.0734	3.0988	3.1242	3.1496	3.1750	3.2004	3.2258	3.2512	3.2766
130	3.3020	3.3274	3.3528	3.3782	3.4036	3.4290	3.4544	3.4798	3.5052	3.5306
140	3.5560	3.5814	3.6068	3.6322	3.6576	3.6830	3.7084	3.7338	3.7592	3.7846
150	3,8100	3.8354	3.8608	3.8862	3.9116	3.9370	3.9624	3.9878	4.0132	4.0386
160	4.0640	4.0894	4.1148	4.1402	4.1656	4.1910	4.2164	4.2418	4.2672	4.2926
170	4.3180	4.3434	4.3688	4.3942	4.4196	4.4450	4.4704	4.4958	4.5121	4.5466
180	4.5720	4.5974	4.6228	4.6482	4.6736	4.6990	4.7244	4.7498	4.7752	4.8006
190	4.8260	4.8514	4.8768	4.9022	4.9276	4.9530	4.9784	5.0038	5.0292	5.0546
200	5.0800	5.1054	5.1308	5.1562	5.1816	5.2070	5.2324	5.2578	5.2832	5.3086
210	5.3340	5.3594	5.3848	5.4102	5.4356	5.4610	5.4864	5.5118	5.5372	5.5626
220	5.5880	5.6134	5.6388	5.6642	5.6896	5.7150	5.7404	5.7658	5.7912	5.8166
230	5.8420	5.8674	5.8928	5.9182	5.9436	5.9690	5.9944	6.0198	6.0452	6.0706
240	6.0960	6.1214	6.1468	6.1722	6.1976	6.2230	6.2484	6.2738	6.2992	6.3246
250	6.3500	6.3754	6.4008	6.4262	6.4516	6.4770	6.5024	6.5278	6.5532	6.5786
260	6.6040	6.6294	6.6548	6.6802	6.7056	6.7310	6.7564	6.7818	6.8072	6.8326
270	6.8580	6.8834	6.9088	6.9342	6.9596	6.9850	7.0104	7.0358	7.0612	7.0866
280	7.1120	7.1374	7.1628	7.1882	7.2136	7.2390	7.2644	7.2898	7.3152	7.3406
290	7.3660	7.3914	7.4168	7.4422	7.4676	7.4930	7.5184	7.5483	7.5692	7.5946
300	7.6200	7.6454	7.6708	7.6962	7.7216	7.7470	7.7724	7.7978	7.8232	7.8486
310	7.8740	7.8994	7.9248	7.9502	7.9756	8.0010	8.0264	8.0518	8.0772	8.1026
320	8.1280	8.1534	8.1788	8.2042	8.2296	8.2550	8.2804	8.3058	8.3312	8.3566
330	8.3820	8.4074	8.4328	8.4582	8.4836	8.5090	8.5344	8.5598	8.5852	8.6106
340	8.6360	8.6614	8.6868	8.7122	8.7376	8.7630	8.7884	8.8138	8.8392	8.8646
350	8.8900	8.9154	8.9408	8.9662	8.9916	9.0170	9.0424	9.0678	9.0932	9.1186
360	9.1440	9.1694	9.1948	9.2202	9.2456	9.2710	9.2964	9.3218	9.3472	9.3726
370	9.3980	9.4234	9.4488	9.4742	9.4996	9.5250	9.5504	9.5758	9.6012	9.6266
380	9.6520	9.6774	9.7028	9.7282	9.7536	9.7790	9.8044	9.8298	9.8552	9.8806
390	9.9060	9.9314	9.9568	9.9822	10.0076	10.0330	10.0584	10.0838	10.1092	10.1346

* 1 Mil = 0.001 inch.

hbar to tonf/in², MN/m², lbf in² and kg/mm²

h bar	tonf/in²	MN/m² N/mm²	lbf/in²	kgf/mm²	hbar	tonf/in²	MN/m² N/mm²	lbf/in²	kgf/mm²
0.5	0.3	5	700	0.5	30.5	19.7	305	44200	31.1
1	0.6	10	1500	1.0	31	20.1	310	45000	31.6
1.5	1.0	15	2200	1.5	31.5	20.4	315	45700	32.1
2	1.3	20	2900	2.0	32	20.7	320	46400	32.6
2.5	1.6	25	3600	2.5	32.5	21.0	325	47100	33.1
3	1.9	30	4400	3.1	33	21.4	330	47900	33.7
3.5	2.3	35	5100	3.6	33.5	21.7	335	48600	34.2
4	2.6	40	5800	4.1	34	22.0	340	49300	34.7
4.5	2.9	45	6500	4.6	34.5	22.3	345	50000	35.2
5	3.2	50	7300	5.1	35	22.7	350	50800	35.7
5.5	3.6	55	8000	5.6	35.5	23.0	355	51500	36.2
6	3.9	60	8700	6.1	36	23.3	360	52200	36.7
6.5	4.2	65	9400	6.6	36.5	23.6	365	52900	37.2
7	4.5	70	10200	7.1	37	24.0	370	53700	37.7
7.5	4.9	75	10900	7.6	37.5	24.3	375	54400	38.2
8	5.2	80	11600	8.2	38	24.6	380	55100	38.7
8.5	5.5	85	12300	8.7	38.5	24.9	385	55800	39.3
9	5.8	90	13100	9.2	39	25.3	390	56600	39.8
9.5	6.2	95	13800	9.7	39.5	25.6	395	57300	40.3
10	6.5	100	14500	10.2	40	25.9	400	58000	40.8
10.5	6.8	105	15200	10.7	40.5	26.2	405	58700	41.3
11	7.1	110	16000	11.2	41	26.5	410	59500	41.8
11.5	7.4	115	16700	11.7	41.5	26.9	415	60200	42.3
12	7.8	120	17400	12.2	42	27.2	420	60900	42.8
12.5	8.1	125	18100	12.7	42.5	27.5	425	61600	43.3
13	8.4	130	18900	13.3	43	27.8	430	62400	43.8
13.5	8.7	135	19600	13.8	43.5	28.2	435	63100	44.4
14	9.1	140	20300	14.3	44	28.5	440	63800	44.9
14.5	9.4	145	21000	14.8	44.5	28.8	445	64500	45.4
15	9.7	150	21800	15.3	45	29.1	450	65300	45.9
15.5	10.0	155	22500	15.8	45.5	29.5	455	66000	46.4
16	10.4	160	23200	16.3	46	29.8	460	66700	46.9
16.5	10.7	165	23900	16.8	46.5	30.1	465	67400	47.4
17	11.0	170	24700	17.3	47	30.4	470	68200	47.9
17.5	11.3	175	25400	17.8	47.5	30.8	475	68900	48.4
18	11.7	180	26100	18.4	48	31.1	480	69600	48.9
18.5	12.0	185	26800	18.9	48.5	31.4	485	70300	49.5
19	12.3	190	27600	19.4	49	31.7	490	71100	50.0
19.5	12.6	195	28300	19.9	49.5	32.1	495	71800	50.5
20	12.9	200	29000	20.4	50	32.4	500	72500	51.0
20.5	13.3	205	29700	20.9	50.5	32.7	505	73200	51.5
21	13.6	210	30500	21.4	51	33.0	510	74000	52.0
21.5	13.9	215	31200	21.9	51.5	33.3	515	74700	52.5
22	14.2	220	31900	22.4	52	33.7	520	75400	53.0
22.5	14.6	225	32600	22.9	52.5	34.0	525	76100	53.5
23	14.9	230	33400	23.5	53	34.3	530	76900	54.0
23.5	15.2	235	34100	24.0	53.5	34.6	535	77600	54.6
24	15.5	240	34800	24.5	54	35.0	540	78300	55.1
24.5	15.9	245	35500	25.0	54.5	35.3	545	79000	55.6
25	16.2	250	36300	25.5	55	35.6	550	79800	56.7
25.5	16.5	255	37000	26.0	55.5	35.9	555	80500	56.6
26	16.8	260	37700	26.5	56	36.3	560	81200	57.1
26.5	17.2	265	38400	27.0	56.5	36.6	565	81900	57.6
27	17.5	270	39200	27.5	57	36.9	570	82700	58.1
27.5	17.8	275	39900	28.0	57.5	37.2	575	83400	58.6
28	18.1	280	40600	28.6	58	37.6	580	84100	59.1
28.5	18.5	285	41300	29.1	58.5	37.9	585	84800	59.7
29	18.8	290	42100	29.6	59	38.2	590	85600	60.2
29.5	19.1	295	42800	30.1	59.5	38.5	595	86300	60.7
30	19.4	300	43500	30.6	60	38.8	600	87000	61.2

Factors

To convert	Multiply by	To convert	Multiply by
in to mm	25.4	mm to in	0.03937
in² to mm²	645.16	mm² to in²	0.00155
in³ to cm³	16.387	cm³ to in³	0.061024
in⁴ to cm⁴	41.623	cm⁴ to in⁴	0.024025
ft to m	0.3048	m to ft	3.2808
ft² to m²	0.092903	m² to ft²	10.764
ft³ to m³	0.028317	m³ to ft³	35.315
yd to m	0.9144	m to yd	1.0936
yd² to m²	0.83613	m² to yd²	1.1960
yd³ to m³	0.76456	m³ to yd³	1.3080
lb to kg	0.45359	kg to lb	2.2046
cwt to kg	50.802	kg to cwt	0.019684
tons to kg	1016.1	kg to tons	0.00098421
tons to lb	2240	lb to tons	0.0004464
tons to short tons	1.12	short tons to tons	0.8929
tons to tonnes (metric)	1.0160	tonnes (metric) to tons	0.98421
lb/in² to kg/mm²	0.0007031	kg/mm² to lb/in²	1422.33
lb/in² to kg/cm²	0.07031	kg/cm² to lb/in²	14.2233
lb/in² to tons/in²	0.0004464	tons/in² to lb/in²	2240
lb/in³ to g/cm³	27.680	g/cm³ to lb/in³	0.036127
lb/ft to kg/m	1.4882	kg/m to lb/ft	0.67197
lb/ft² to kg/m²	4.8824	kg/m² to lb/ft²	0.2048
tonf/in² to MN/m² or MPa or N/mm²	15.444	MPa or MN/m² to tonf/in²	0.064749
tonf/in² to hbar	1.5444	hbar to tonf/in²	0.64749
tonf/in² to kgf/mm²	1.5749	kgf/mm² to tonf/in²	0.63497
lbf/ft² to N/m² or Pa	47.880	Pa or N/m² to lbf/ft²	0.02089
lbf/in² to N/m² or Pa	6894.8	Pa or N/m² to lbf/in²	0.00014504 .
lbf/in² to hbar or Pa	0.00068948	hbar to lbf/in²	1450.4
kgf/m² to N/m² or Pa	9.8067	Pa or N/m² to kgf/m²	0.10197
kgf/mm² to hbar	0.98067	hbar to kgf/mm²	1.0197
cal cm/cm² s °C to W/m °C	418.68	W/m °C to cal cm/cm² s °C	0.0023885
$\mu\Omega$ m to Ω m	10^{-6}	Ω m to $\mu\Omega$ cm	10^{8}

Other conversions

	Multiply by
Density	
Pounds/cubic inch to kilograms/cubic metre	27680
Pounds/cubic foot to kilograms/cubic metre	16.018
Tons/cubic yard to kilograms/cubic metre	1328.9
Force	
Poundals to newtons	0.13825
Pounds force to newtons	4.448
Tons force to newtons	9964

Torque

Pounds force-inch to newton-metres	0.11298
Pounds force-feet to newton-metres	1.3558
Tons force-feet to newton-metres	3037
Inches of mercury to millibars	33.864
Inches of mercury to newtons per square metre or pascal	3386.4
Note. 1 bar = 10^5 newtons per square metre or 10 Pa	

Work, Energy

Therms to mega-joules	105.5
Kilowatt-hours to mega-joules	3.6
British thermal unit to joules (metre-newtons)	1055.1
Centigrade heat unit to joules (metre-newtons)	1899.2
Foot pound f. to joules (metre-newtons)	1.3558

Power

Horse-power to watts (joules per sec)	745.7
Foot pound f. per sec to watts (joules per sec)	1.3558

Illumination

Lumens per square foot ⎱ to lumens per square	
Foot candles ⎰ metre (lux)	10.764

Angular measurement

Radians to degree	57.29
Degrees to radians	0.01745

2   Selection of British Standards relating to welding

Note. When consulting British Standards the engineer should make sure that the Standard concerned incorporates the latest amendments.

See also notes on acceptance levels for defects in fusion welded joints, PD 6493

	BS number
Welding terms and symbols:	
Part 1 Welding, brazing and thermal cutting glossary	
Part 2 Symbols for welding	
Part 3 Terminology of and abbreviations for fusion weld imperfections as revealed by radiography	*499*
Arc welding plant, equipment and accessories	*638*
Approval testing of welders when welding procedure is not required	*4872*
Tests for use in the training of welders; manual metal arc and oxy-acetylene welding of mild steel	*1295*
Covered electrodes for the manual metal arc welding or carbon and carbon–manganese steels	*639*
Low-alloy steel electrodes for manual metal arc welding	*2493*
Chromium nickel austenitic and chromium steel electrodes for manual arc welding	*2926*
Class I welding of ferritic steel pipework for carrying fluids	*2633*
Class I arc welding of stainless steel pipework for carrying fluids	*4677*
Class II metal arc welding of steel pipelines and pipe and pipe assemblies for carrying fluids	*2971*

Shell boilers of welded construction (other than water-tube boilers):
   Part 1 Class I welded construction
   Part 2 Class II and Class III welded construction   2790

Vertical cylindrical welded steel storage tanks for low temperature service.
Single-wall tanks for temperature down to − 50°C.   4741

Transportable gas containers, steel containers up to 130 litres water
capacity with automatic all welded circumferential seams   5045

Fusion welded pressure vessels for general purposes: 1500
   Part 3 Aluminium
        Carbon and low alloy steels   1500A

Fusion welded pressure vessels for use in the chemical, petroleum and allied
industries:
   Part 1 Carbon and ferritic alloy steels
   Part 2 Austenitic stainless steels   1515

Projection welding of low carbon steel sheet and strip   2630

General requirements for seam welding in mild steel   2937

Identification of contents of industrial gas containers   349

Safety

Equipment for eye, face and neck protection against radiation arising
during welding and similar operations   1542

Filters for use during welding and similar industrial operations   679

Protective clothing for welders   2653

Flameproof industrial clothing (materials and design)   1547

Industrial gloves   1651

Safety footwear   1870

Industrial overalls for general purposes   1907

Industrial eye protectors   2092

Safety colours for use in industry   2929

Testing

Crack tip opening displacement testing (CTOD – previously COD)   5762

Methods of testing fusion welds in copper and copper alloys   4206

Methods for radiographic examinations of fusion welded circumferential
butt joints in steel pipes   2910

Methods of radiographic examination of fusion welded butt joints in steel   2600

Methods of testing fusion welds in aluminium and aluminium alloys   3451

Methods of testing fusion welds in copper and copper alloys   4206

Methods of testing fusion welded joints and weld metal in steel   709

Methods for magnetic particle testing of welds   4397

Methods for penetrant testing of welded or brazed joints in metals   4416

Methods for ultrasonic examination of welds:
   Part 1 Manual examination of fusion welded butt joints in ferritic steels
   Part 2 Automatic examination of fusion welded butt joints in ferritic
       steels   3923

Spot and seam welding, etc.

Gas welding

3   *Notes on Published Document (PD) 6493 of the British Standards I: Guidance on some methods for the derivation of acceptance levels for defects in fusion welded joints*

Choice of parent plate and weld metal can be made after tensile, compression, fatigue, impact and hardness etc. tests are made and, when a welded structure or component is completed, NDT tests (X- or gamma-ray, magnetic, ultrasonic etc.) are employed to determine the presence of various defects which may cause failure in service. These defects may be divided into planar and non-planar:

Planar defects
- (*a*) cracks
- (*b*) lack of fusion or penetration
- (*c*) undercut, root undercut or concavity and overlap

Non-planar defects
- (*a*) cavities
- (*b*) solid inclusions

Methods of failure
- (*a*) brittle fracture
- (*b*) fatigue
- (*c*) overload of the remaining cross-section
- (*d*) leakage in vessels
- (*e*) corrosion, erosion, corrosion fatigue, stress corrosion
- (*f*) buckling
- (*g*) creep or interaction between creep and fatigue

From a study of PD 6493 it will be seen that some of these defects may be acceptable whilst others are not. These latter must be rectified otherwise there is a risk of failure. All of this is subject to agreement between interested parties.

If defects do not exceed quality control level, there need not be any further action.

If acceptance levels have been agreed upon based on the Engineering Critical Assessment (ECA) for materials used, welding consumables and processes, welding procedure and any stress factors due to wind, water etc. defects should be assessed on this base.

If no acceptance levels have been agreed upon, an ECA based on the guidance given in PD 6493 or an agreed alternative should be carried out.

The student can obtain much guidance from this document and can see clearly the amount of elastic–plastic fracture mechanics involved. The applications are complicated and the use of PD 6493 should be subject to expert advice.

Note on *Crack tip opening displacement testing* (CTOD) BS 5762*

Students requiring more information on this on-going subject must refer to BS 5762 on CTOD testing, which gives full details of the many variables concerned. The various articles in *Metal Construction* (the journal of the Institute of Welding) and research papers on the subject should be consulted. The fracture laboratory of the Institute is well equipped with the latest machines and undertakes the CTOD testing of specimens and weldments for industry and provides exhaustive test reports on them.

---

* This test is now called crack tip opening displacement testing (Amendment 4131 1982) to avoid confusion with other displacements within the crack.

City and Guilds of London Institute examination questions

Welding science, metallurgy and technology

Note: All dimensions are given in millimetres, unless otherwise stated.

1 A butt weld is to be made by the manual metal arc process between two 2 m by 1 m by 10 mm thick low-carbon steel plates along the long side.

   (a) Explain briefly why the cooling rate should be controlled.

   (b) State *three* factors which may influence the cooling rate of the weld.

2 What is meant by distortion of welded work? State *four* factors which may cause distortion during the welding of mild steel assemblies. Describe briefly one method used to control distortion when building up a short section of a 75 mm diameter steel shaft worn below the correct diameter.

3 Discuss briefly safety recommendations with regard to each of the following:

   (a) the type of current used in dangerous situations,

   (b) metal arc welding in confined spaces,

   (c) effects due to (1) arc radiations and (2) heat exhaustion,

   (d) earthing and conductivity of the welding return circuit.

4 Give *one* reason for the use of *each* of the following in welded work:

   (a) a chill,

   (b) a heat retaining material,

   (c) a fixture.

5 Make sectional sketches of the following weld joints and give their weld symbol in accordance with the appropriate BS 499:

   (a) close-square-tee fillet (weld both sides),

   (b) single bevel butt.

6 What are meant by the following terms:
   (*a*) conduction,
   (*b*) convection,
   (*c*) radiation.
   Give *one* example of each welding practice.

7 Explain the procedure which must be carried out in order to make an effective macroscopic examination of a transverse section through a welded joint in low-carbon steel, naming *four* defects which may be revealed by this method of examination.

8 A dye penetrant method may be used for detecting defects in a welded joint.
   (*a*) Outline the principles of this method.
   (*b*) What type of defects may be revealed?

9 Explain briefly the meaning of *each* of the following electrical terms:
   (*a*) voltage,
   (*b*) current,
   (*c*) resistance.

10 State *two* advantages in *each* case of:
   (*a*) hot working,
   (*b*) cold working in a low-carbon steel.

11 (*a*) Describe briefly the effect of cold rolling on the grain structure of a metal.
   (*b*) Explain what takes place when a metal that has been cold rolled is heated to its recrystallization temperature.

12 Describe briefly the difference between the current pick-up systems for the generation of alternating current and direct current.

13 (*a*) Give *two* reasons why grain growth may take place in a metal.
   (*b*) What effect will enlarged grain structures have on the mechanical properties of a metal?

14 Explain briefly why oxygen and nitrogen should be excluded throughout the welding operation.

15 In relation to manual metal arc welding state *two* functions in *each* case of:
   (*a*) fluxes,
   (*b*) slags.

16 For each of the following cases state which kind of cracking is most likely to occur in a fusion welded joint:
   (*a*) a weld highly stressed during the early stages of solidification,
   (*b*) a weld in a hardenable steel made without pre-heat,
   (*c*) a weld in an unstabilized austenitic stainless steel.

17 In the welding of a solution-treatable type of aluminium alloy, describe any *two* weldability difficulties that you would expect to encounter.

18   (*a*) Name two practical difficulties likely to be encountered in inspecting a weld joint by radiographic means.

(*b*) Why are magnetic crack detection methods not used for examining welds in copper alloys?

(*c*) What crack detection method could be used for copper alloys?

19   (*a*) For a plain carbon steel containing 0.4%C list three typical metallurgical states which might exist in the material in the vicinity of a fusion weld.

(*b*) For each of the conditions under (*a*), outline the sequence of heating and cooling that would put the material in that particular condition.

20 State the purpose of a drooping characteristic for arc welding. Sketch the form of a drooping characteristic and indicate on it:

(*a*) the open circuit voltage,

(*b*) the average arc voltage,

(*c*) the average welding current.

21 On a labelled outline sketch:

(*a*) name the different types of structure that you would expect to find in the vicinity of a single-run-vee butt weld made in an initially *annealed* solution-heat-treatable alloy, and

(*b*) indicate the approximate areas in which you would expect to find each structure.

Note: you are expected only to name the type of structure, not to show any details.

22   (*a*) Why is it that residual stress tends to become less of a problem the faster you are able to complete an arc-welded joint?

(*b*) Give *one* reason why a particular material might be very liable to hot cracking.

23 State briefly any *two* problems likely to be met in trying to weld an alloy containing one relatively low-melting-temperature constituent and with a wide solidification range of temperature. What is meant by the term 'low-alloy steel'?

24   (*a*) What is the most commonly used ferrous alloy?

(*b*) Name *two* different non-ferrous alloys used in welded fabrications.

(*c*) State *two* conditions essential for the formation of an equi-axial crystal structure, in an arc welded deposit in the as-welded condition.

25   (*a*) In *each* of the following cases state a possible cause of *one* type of cracking which may be produced in fusion welded joints made in:
      (1) low-alloy hardenable steels,
      (2) austenitic heat resisting steels.
  (*b*) Outline why it is important to avoid sharp corners in fusion welded joints made in structural steel for service at low temperatures.

26   (*a*) What is meant by the critical cooling rate of a plain carbon steel?
  (*b*) State *two* detrimental effects which may be produced during the making of a welded joint in 0.4% carbon steel if the critical cooling rate is exceeded.

27   (*a*) Describe the effects on the weldability of low-alloy steel of any *two* of the following elements:
      (1) nickel,
      (2) chromium,
      (3) hydrogen.
  (*b*) Explain how the difficulties which may arise from the presence of hydrogen may be overcome during the welding of low-alloy steels.

28 A low-carbon steel open-top tank 7 m in mean diameter by 5 m deep by 20 mm thick is to be fabricated by welding. Calculate the total mass of plate if the metal density is 7830 kg/m^3, taking π as 22/7.
Ten low-carbon steel plates each 3 m by 2 m by 20 mm thick are required to make an oil tank. Calculate the total mass of plate used if the metal density is 7830 kg/m^3.

29 Make a sectional sketch of *each* of the following types of welded joint, giving the appropriate weld symbol to show these joints in accordance with BS 499:
  (*a*) single 'U' butt,
  (*b*) double-bevel butt.

30 Describe with the aid of sketches where appropriate, how defects may be detected by using each of the following methods of testing:
  (*a*) X-ray,
  (*b*) ultrasonic,
  (*c*) dye penetrant.

31 Explain by means of a sketch what is meant by temperature gradient.

32 State two important functions of fluxes used during oxy-acetylene welding operations.

33 Name two main types of weld testing and give one example of each type.

34 Stainless steel filler rod used for oxy-acetylene welding usually contains an element known as a stabilizer. Name one such element and state its main purpose.

35　(*a*)　What is meant by 'post-heating'?

　　(*b*)　State two reasons why welded assemblies may be subjected to post-heating.

36 Increasing the carbon content of a plain steel influences certain physical properties. Using one word only in each case, state the effect of increase in carbon content on the following properties:

　　(*a*)　tensile strength,

　　(*b*)　elongation,

　　(*c*)　hardness,

　　(*d*)　melting points.

37　(*a*)　Describe, in detail, the preparation and procedure for making a controlled root bend test from a 5 mm thick low-carbon steel test piece taken from a butt welded joint.

　　(*b*)　State two desirable features such a bend test should reveal.

38 State briefly how any *four* of the following may arise in welding practice and explain how *each* may be counteracted:

　　(*a*)　grain growth in a brass,

　　(*b*)　over-ageing of precipitation hardenable aluminium alloys,

　　(*c*)　residual stresses in low-carbon steel,

　　(*d*)　interangular corrosion of austenitic stainless steel,

　　(*e*)　cold cracking of low-alloy, high-tensile steel.

39 Outline *two* workshop methods of distinguishing a grey iron casting from a malleable iron casting.

40 Give two examples of difficulties which may be encountered in the welding of high thermal conductivity materials.

41 Describe with the aid of a simple sketch what is meant by a 'pearlitic structure'.

42　(*a*)　What is meant by 'hot shortness'?

　　(*b*)　Give two causes of this weakness.

43 Why is alternating current potentially more dangerous than direct current at the same nominal voltage.

44　(*a*)　In what type of weld would you expect to find a columnar structure?

　　(*b*)　With the aid of a sketch, show in which part of the weld you would find this structure.

45 Explain how rapid cooling affects the microstructure of an 18% nickel-steel weld metal.

46 Explain the essential difference between macroscopic and microscopic examination.

47 The figure shows a pictorial view of a bracket to be produced by welding. Sketch, approximately half full size, an elevation in direction of arrow '*A*' and a plan view in direction of arrow '*V*'. Insert on the sketch the appropriate weld symbols according to BS 499 in order to indicate the welded joints necessary to fabricate the bracket.

48   (*a*) What is meant by a metallic solid solution?
      (*b*) Give one example of a solid solution using an alloy which is normally welded.

Question 47

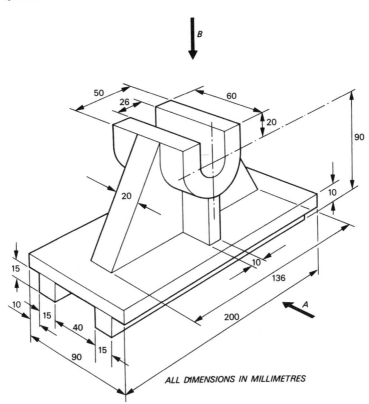

ALL DIMENSIONS IN MILLIMETRES

49 State the purpose of each of the following components of an arc welding plant:
  (*a*) rectifier,
  (*b*) transformer,
  (*c*) choke reactance.

50 Briefly explain the function of *each* of the following in manual metal arc welding:
  (*a*) a low-voltage safety device,
  (*b*) a rectifier.

51 List *six* methods of testing welded joints, indicating clearly whether the methods are destructive or non-destructive.

52   (*a*) Give *three* reasons why pre-heating is sometimes necessary when manual metal arc welding.
  (*b*) Briefly describe how pre-heating may be carried out by an electro-thermal method.

53 State *two* physical properties and *one* metallurgical problem that must be encountered when joining dissimilar materials by a fusion welding process.

54 In *each* case, list *six* factors that would require attention while carrying out visual inspection of metal arc welded work:
  (*a*) prior to welding,
  (*b*) during welding,
  (*c*) after welding.

55 Describe the effects on the weldability of steel of *each* of the following elements and state in *each* case *one* method of overcoming difficulties that may arise:
  (*a*) carbon,
  (*b*) hydrogen,
  (*c*) silicon,
  (*d*) chromium.

56   (*a*) Explain why a destructive macrostructural examination of the cross-section of a weld is often required.
  (*b*) Why is the welding craftsman not normally expected to make a macrostructural examination of a weld?

57   (*a*) Why may magnetic particle inspection only be used for crack detection in carbon steel and some alloy steels, whereas dye penetrant methods can be applied to all metals?
  (*b*) Name *four* of the most possible causes of cracking in welds made by metal arc gas-shielded welding.

58 The figure shows the dimensions of a gusset plate. Calculate the area of plate required to make 50, assuming no wastage.

59 Draw and fully dimension, in the direction of arrow *A*, an end elevation of the welded bracket shown in the figure. Insert the appropriate weld symbols according to BS 499 on the welded joints used.

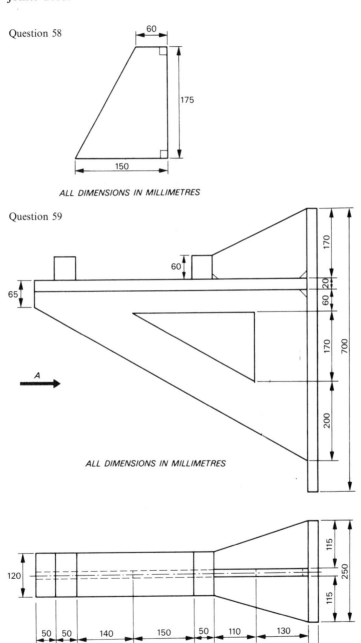

Question 58

ALL DIMENSIONS IN MILLIMETRES

Question 59

ALL DIMENSIONS IN MILLIMETRES

60 Give three reasons for the use of pre-heat in welding applications. Explain one method of pre-heat.

61 (*a*) Give two causes of hot cracking in carbon-steel fusion welds.

    (*b*) Name one impurity that may cause hot cracking in welded steel joints.

    (*c*) State one way in which hot cracking can be minimized by welding procedure.

62 (*a*) Explain briefly what is meant by solution treatment.

    (*b*) Give one example of an alloy that may be solution treated.

    (*c*) What effect will fusion welding have on the mechanical properties of a solution-treated alloy?

63 State *two* conditions essential for the formation of an equi-axial crystal structure, in an arc welded deposit in the as-welded condition.

64 If treated alloy plates, in the fully solution-treated and aged condition, are joined by fusion welding state whether:

    (*a*) the as-welded deposit will be harder or softer than the parent plate,

    (*b*) the heat-affected zone will be harder or softer than the parent plate.

65 (*a*) Briefly describe the mode of solidification leading to columnar grain structure in an autogenous welded joint made by tungsten-arc welding.

    (*b*) Give *one* example of a type of material in which a band of refined grain structure may be expected nearest to the weld boundary in the heat-affected zone of a fusion welded joint.

66 (*a*) Give *two* examples of when pre-heating is essential in the fusion welding of carbon steels.

    (*b*) Explain why the presence of moisture in any form should be avoided when gas-shielded arc welding low-alloy steels.

67 (*a*) What is meant by the critical cooling rate of a plain carbon steel?

    (*b*) State *two* detrimental effects which may be produced during the making of a welded joint in 0.4% carbon steel if the critical cooling rate is exceeded.

68 (*a*) Give *one* example of *either* dilution *or* pick-up effects arising in gas-shielded arc welding practice.

    (*b*) State the shielding gas or gas mixture which is best suited to obtain the required modes of metal transfer for the effective

metal arc gas-shielded welding of *each* of the following:
   (1) low-carbon steel by dip transfer,
   (2) austenitic stainless steel by controlled spray (pulse) transfer.

69   (a) Explain, with the aid of a sketch, how the level of dilution of a butt-welded joint may be determined.

   (b) Show by means of a labelled sketch *one* type of edge preparation used to control pick-up effects when making a butt-welded joint in clad steel.

70 State what is meant by a metallic alloy and name one non-ferrous alloy used in welded fabrications.

71 Give *three* reasons why hot cracking may be a problem when manual metal-arc welding austenitic stainless steel.

72   (a) Explain the difference between stress relieving and annealing in the heat treatment of steels.

   (b) State whether the heat-affected zone will be harder or softer than the parent plate in a fusion welded joint made in fully solution-treated and aged alloy plates.

73   (a) The figure shows the arrow line and reference line according to BS 499 Part 2, 1980. State for both the symbols shown what information they convey.

   (b) State what is meant by the critical cooling rate of a plain carbon steel.

   (c) Give *two* examples of when pre-heating is essential in the fusion welding of carbon steels.

   (d) List *four* factors that should be considered when determining the pre-heating temperature to be used for welding a steel fabrication.

74   (a) Explain the difference between the heat treatment processes 'annealing' and 'normalizing'.

   (b) State *three* advantages obtained by normalizing alloy steel welded joints.

75 The equivalent carbon content of an alloy steel can be found from the formula:

$$\text{Carbon equivalent} = \%\text{C} + \frac{\text{Mn}}{6} + \frac{\text{Cr} + \text{Mo} + \text{V}}{5} + \frac{\text{Ni} + \text{Cu}}{15}$$

Question 73

The composition of an alloy steel is as shown in the table:

Carbon	Phosphorus	Sulphur	Vanadium	Chromium	Manganese	Silicon
0.22%	0.05%	0.05%	0.10%	0.10%	1.50%	0.50%

    (*a*) Using the formula given, calculate the carbon equivalent.

    (*b*) State *two* precautions to be taken when welding this type of steel.

76 State what is meant by a metallic alloy and name one non-ferrous alloy used in welded fabrications.

77 If the parent metal composition of a welded joint in plain carbon steel is 0.22% carbon, the all-weld metal deposit composition is 0.11% carbon and the cross-sectional area of the weld metal zone is 10 times the size of the cross-sectional area of the fusion zone, estimate the approximate average carbon content of the weld deposit resulting from dilution.

78   (*a*) Give one reason why the hard brittle form of structure (martensite) is most likely to form close beside the fusion boundary in the heat-affected zone of a welded joint in a hardenable steel.

        (*b*) Name the kind of structure to be found *just outside* a martensitic zone in the heat-affected zone of a welded joint in a hardenable steel (*Note*. If you do not know the technical name of the structure a simple word description will do.)

79   (*a*) What is the purpose of a rectifier when used for welding from a.c. power supply?

        (*b*) On a simple labelled graph shown clearly the typical form of current flow likely to be obtained from a welding rectifier.

80 A metallic alloy may have a 'narrow' or a 'wide' solidification range.

        (*a*) State which type of solidification mode will give most difficulty in fusion welding.

        (*b*) Give *one* reason to justify your answer.

81   (*a*) What is the most commonly used ferrous alloy?

        (*b*) Name *two* different non-ferrous alloys used in welded fabrications.

82   (*a*) If your welding generator caught fire and you could not switch off the supply current, what type of fire extinguisher would you use?

        (*b*) Is there any type of extinguisher that you should not use?

        (*c*) Why should you not use the type of extinguisher in (*b*).

83 Explain why a welding generator neither blows its fuse nor burns out when a short-circuit occurs at the welding electrode.

84 In the welding of a solution-treatable type of aluminium alloy, describe any *two* weldability difficulties that you would expect to encounter.

85 If you are *tungsten arc* gas-shielded welding a butt joint in 0.4% C steel plate 12 mm thick with 0.1% C steel filler wire, estimate the approximate average carbon content of the deposit.

86 If you are *metal arc* gas-shielded welding a butt joint in 0.4% C steel plate 12 mm thick with 0.1% C steel filler wire, estimate the approximate average carbon content.

87 (*a*) With what type of material would you expect to find equi-axial solidification occurring in a fusion welded deposit?

   (*b*) Why is a columnar growth almost invariably found in the structure of a progressive fusion weld in the as-welded state?

   (*c*) State the type of grain structure which may be found in the heat-affected zone of a single-run weld made in normalized low-carbon steel.

88 In the cross-sectional shape of a fusion welded joint, sharp corners should be avoided.

   (*a*) Give the most important reason for this precaution.

   (*b*) State why this precaution is particularly important when welding structural steel for service in a cold atmosphere.

89 When would it not be safe to connect two welding generators in parallel to give increased power to a single arc?

90 (*a*) Why is it that residual stress tends to become less of a problem the faster you are able to complete an arc-welded joint.

   (*b*) Give *one* reason why a particular material might be very liable to hot intergranular cracking during fusion welding.

91 (*a*) What is dilution in fusion welding?

   (*b*) What is pick-up in fusion welding?

   (*c*) Can the atmosphere surrounding an arc affect any pick-up that normally tends to occur?

92 Draw a simple outline sketch of the cross-section of a two-run double butt weld in a hardenable steel made without pre-heating and show (*a*) *three* different types of structure that might be found in the heat-affected zones, and (*b*) the most likely location(s) of each of the types you give. You are not expected to

give details of the structure; a simple general word description will be sufficient if you do not know the technical name of a particular structure.

93 State briefly any *two* problems likely to be met in trying to weld an alloy containing one relatively low-melting-temperature constituent and with a wide solidification range of temperature.
What is meant by the term 'low-alloy steel'?

94 State briefly how any *four* of the following may arise in welding practice and explain how *each* may be counteracted:
  (*a*) grain growth in brass,
  (*b*) over-ageing of precipitation hardenable aluminium alloys,
  (*c*) residual stresses in low-carbon steel,
  (*d*) intergranular corrosion of austenitic stainless steel,
  (*e*) cold cracking of low-alloy, high-tensile steel.

95 State three ways in which weather conditions may adversely affect welding operations.

96 (*a*) In what form would you expect the carbon to be present in (1) white cast iron, (2) grey cast iron.
  (*b*) Which of these types of iron would most likely be formed in the heat-affected zone if the cooling rate after welding was too fast.

97 (*a*) Give any three advantages obtained when using rectifier welding equipment.
  (*b*) Explain what is meant by the terms 'arc voltage' and 'open circuit voltage'.
  (*c*) Give three probable causes of poor-quality resistance spot welds.

98 (*a*) Name *three* obnoxious fumes or poisonous gases which may be formed during metal arc welding operations.
  (*b*) Give *two* safety precautions to be taken in order to avoid personal injury from these fumes or gases.

99 List *six* methods of testing welded joints, indicating clearly whether the methods are destructive or non-destructive.

100 (*a*) Explain what is meant by dilution in weld deposits.
  (*b*) List *three* factors which may influence the amount of dilution produced in a weld deposit.

101 (*a*) Briefly explain why notch effects must be avoided in stressed welded structures.
  (*b*) Sketch *two* defects and *two* undesirable weld contours, *each* of which could create notch effects.

102   (*a*) For a plain carbon steel containing 0.4% C list three typical metallurgical states which might exist in the material in the vicinity of a fusion weld.

     (*b*) For each of the conditions under (*a*), outline the sequence of heating and cooling that would put the material in that particular condition.

103 For each of the following cases state which kind of cracking is most likely to occur in a fusion welded joint:

     (*a*) a weld highly stressed during the early stages of solidification,

     (*b*) a weld in a hardenable steel made without pre-heat,

     (*c*) a weld in an unstabilized austenitic stainless steel.

104   (*a*) Name two practical difficulties likely to be encountered in inspecting a weld joint by radiographic means.

     (*b*) Why are magnetic crack detection methods not used for examining welds in copper alloys?

     (*c*) What crack detection method could be used for copper alloys?

105 Make a pictorial sketch of the bracket shown in the figure.

Question 105

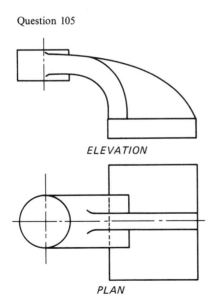

ELEVATION

PLAN

106 With the aid of sketches, explain what is meant by *each* of the following weld sequences:
  (*a*) skip,
  (*b*) block,
  (*c*) back step.

107 A butt welded specimen is to be tested for impact value.
  (*a*) Name a test that could be used.
  (*b*) State the mechanical property that would be measured.
  (*c*) State the effect that low temperature has on the impact resistance of carbon steels.

108 Explain the differences between brittle fracture and hot cracking which may occur in welded fabrications. In *each* case give *one* reason why these types of failure may occur.

109 The figure shows a steel plate with *three* slots cut in it. Calculate the surface area remaining.

Take π as $\dfrac{22}{7}$.

110 Martensite may be formed when arc welding steel.
  (*a*) State *two* conditions which may cause it to be formed.
  (*b*) State *three* methods which could be used to prevent its formation.

111 Steels may be classified according to their range of tensile strengths expressed in newtons per square millimetre.
  (*a*) Explain what is meant by the term newton.
  (*b*) Name and explain the test shown in the figure.

112 In the table below each term used in column *A* is directly related to *one* of the terms used in column *B*. Pair each of the terms listed in column *A* with the appropriate term in column *B*.

Column *A*	Column *B*
Hot cracking	High arc voltage
Health hazard	Iron sulphide
Cellulose covering	Phosgene gas
Rutile covering	High carbon equivalent
Cold cracking	Titanium dioxide

Question 109

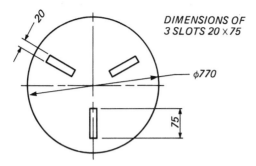

DIMENSIONS OF
3 SLOTS 20 × 75

ⴼ770

20

75

Question 111

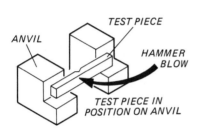

ANVIL

TEST PIECE

HAMMER
BLOW

TEST PIECE IN
POSITION ON ANVIL

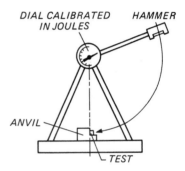

DIAL CALIBRATED
IN JOULES   HAMMER

ANVIL

TEST

Index